Katrin Sedlmeier

Near future changes of compound extreme events from an ensemble of regional climate simulations

Wissenschaftliche Berichte des Instituts für Meteorologie und Klimaforschung des Karlsruher Instituts für Technologie (KIT)
Band 82

Herausgeber: Prof. Dr. Ch. Kottmeier

Institut für Meteorologie und Klimaforschung
am Karlsruher Institut für Technologie (KIT)
Kaiserstr. 12, 76128 Karlsruhe

Eine Übersicht aller bisher in dieser Schriftenreihe erschienenen Bände finden Sie am Ende des Buches.

Near future changes of compound extreme events from an ensemble of regional climate simulations

by
Katrin Sedlmeier

KIT Scientific Publishing

Karlsruher Institut für Technologie
Institut für Meteorologie und Klimaforschung

Near future changes of compound extreme events from an ensemble of regional climate simulations

Zur Erlangung des akademischen Grades eines Doktors der Naturwissenschaften von der KIT-Fakultät für Physik des Karlsruher Instituts für Technologie (KIT) genehmigte Dissertation

von Dipl.-Phys. Katrin Sedlmeier

Tag der mündlichen Prüfung: 30. Oktober 2015
Referent: Prof. Dr. Christoph Kottmeier
Korreferent: PD Dr. Michael Kunz

Impressum

KIT Scientific Publishing

Karlsruher Institut für Technologie (KIT)
KIT Scientific Publishing
Straße am Forum 2
D-76131 Karlsruhe

KIT Scientific Publishing is a registered trademark of Karlsruhe Institute of Technology.
Reprint using the book cover is not allowed.

www.ksp.kit.edu

Print on Demand 2019 – Gedruckt auf FSC-zertifiziertem Papier

ISSN 0179-5619
ISBN 978-3-7315-0476-4
DOI 10.5445/KSP/1000051732

Abstract

The simultaneous occurrence of two or more extreme events, e.g. a heat wave and a drought, also referred to as a compound extreme event, has a low probability but a possibly higher impact than a univariate extreme event. Therefore, high resolution regional information about future changes in their number and temporal succession as well as the uncertainty of these changes are of high relevance when it comes to planning of climate change related adaptation and mitigation measures. However, the methods to analyze compound extreme events are by far not as manifold and established as for their univariate counterparts and regional information about their near future changes is rare.

This thesis aims to help fill this gap by elaborating high resolution regional information about the near future changes of compound hot and dry extreme events in summer and compound cold and wet extreme events in winter for central Europe, including the associated uncertainties. The data basis is a 12 member ensemble of regional climate simulations at the Karlsruhe Institute of Technology (KIT). The ensemble was partly generated within this work using the regional climate model COSMO-CLM at two different resolutions, 50 km (including all of Europe, further referred to as 50 km ensemble) which was then downscaled to 7 km for central Europe (further referred to as 7 km ensemble) and two time periods, one in the recent past (1971-2000, reference period) and one in the near future (2021-2050). At the time of writing, the 7 km ensemble was the largest existing ensemble at this high resolution for the aforementioned region and time periods.

The compound extreme events were analyzed by using three different methods, the first one focusing on absolute extreme events defined by threshold exceedances and the second one defining extremes as strong deviations relative to the local mean state. A new method was developed within this work which focuses on the dynamical behavior of compound extremes, i.e. the temporal succession and ordering of compound extreme episodes, an aspect which has been mostly neglected up to now but which is highly important for the impact these compound extremes may have.
For validation purposes, the 7 km ensemble statistics were compared to results derived from observational data for the reference period. Furthermore, the added value of the computationally more expensive but higher resolved 7 km with respect to the 50 km ensemble was assessed. In general, the 7 km ensemble is able to reproduce the observed statistics of different aspects of the two kinds of extreme events. It performs better with respect to hot and dry extreme events in summer than for cold and wet extremes in winter. In comparison to the 50 km ensemble, the 7 km ensemble shows added value for mean values and absolute extremes, mostly the spatial correlation of extreme events is improved (the correlation coefficient increases by up to 0.2). For extremes relative to the local mean conditions and the dynamical behavior no added value of the higher resolved ensemble can be found.
The climate change signal between 1971-2000 and 2021-2050 of compound temperature and precipitation extremes was calculated by comparing the results of the two different time periods for the 7 km ensemble. As additional information, the ensemble consistency and significance of the change signal as measures of the robustness of the changes were assessed. The change signal of compound extreme events between the reference period and the near future depends strongly on the type of extreme, the aspect of the extreme considered (by the three different methods) and the region. The number of absolute compound hot and dry extreme events increases in all of the investigation area (averaged over the whole investigation area the relative increase is higher than 100%), the absolute changes are strongest in regions

which were already most affected during the reference period like the Rhine Valley (up to 7.5 days per year). Extremes relative to the local mean state show robust changes in the Alps, Bavaria and the Czech Republic of around 100% . In these regions, the dynamical behavior also changes and extreme episodes relative to the local mean state are to be expected more frequently in the near future, the mean time between episodes decreases by 5 to 15 days.Compound cold and wet extreme events only show robust changes in small parts of the investigation area for all three methods. Absolute extremes increase in small parts of the eastern Alps (by about 20%). The analysis of the dynamic properties yields a robust increase in duration in northeastern and northwestern Germany and the time between episodes decreases (by up to 200 days) in parts of western investigation area.

Finally, the relation of different weather types and absolute extreme events was studied and its change signal analyzed. For both kinds of compound extremes, weather types were identified which favor the occurrence of the respective compound extreme events.

Contents

1 Motivation and objectives

The climate of the region we live in influences our lives in many ways, from agriculture, the amount of energy we use, to how we build our houses. Among scientists, there is a broad consensus, that the climate is changing and will change even more in the future. Besides changes in mean values, e.g. an increase in temperature, the occurrence and intensity of extreme events like droughts, floods, heat waves or hurricanes is likely to change (IPCC, 2013). These events have a low probability but a potentially high impact on society. One example is the heat wave of 2003 which highly affected the health of people, agriculture, the ecosystem and infrastructure (e.g., Ciais et al., 2005; Fink et al., 2004). The severity of this event was partly due to a long drought preceding the heat wave (Fischer et al., 2007).

In many cases, extreme events which have a large impact do not only depend on one variable, but are rather of multivariate nature. For example, when heavy precipitation occurs, the flood risk is higher when the soil water content is high, and health issues are more prominent if it is not only very hot but also humid at the same time. Wild fires are more likely to occur if it is hot and dry and their damage potential is enhanced by high wind speeds (Keetch et al., 1968). In the special report on "Managing the Risks of Extreme Events and Disasters to Advance Climate Change Adaptation" by the Intergovernmental Panel on Climate Change, IPCC (IPCC, 2012), these multivariate extremes are referred to as compound extreme events. Three different definitions are given: 1) "two or more extreme events occurring simultaneously or successively", 2) "combinations of extreme events with underlying conditions that amplify the impact of the event", 3) "combinations of events that are not themselves extremes but lead to an extreme event or impact when combined". Although society is highly vulnerable to compound extreme events, they are largely underrepresented in the published literature and yield an interesting and important field of research which has only started growing recently.

The methodological framework for the analysis of compound extreme events is by far less developed than for their univariate counterparts; this is however slowly changing. Recent books on extreme value analysis for climate science usually include an introduction to the framework of multivariate statistics (e.g., Coles et al., 2001; Beirlant et al., 2006) and approaches from finance of risk management, where the analysis of compound events is more common, have been translated to climate science. One possibility of analyzing compound extreme events which perhaps has been applied most frequently, is to use indices that depend on more than one variable. Examples are the wildfire index KBDI (Keetch–Byram drought index, Keetch et al., 1968) which is calculated from temperature, precipitation and wind data or the revised CEI (Climate Extremes Index, Gallant et al., 2014). It depends on temperature, precipitation and soil moisture and can be used to analyze different types of extremes connected to these variables. Some studies analyze joint quantiles of different variables, e.g. Beniston (2009) for temperature and precipitation and Fischer and Knutti (2013) for temperature and humidity. Another approach, which has been used for some time in risk management and finance and is now more widely applied in climate science is the use of copulas (Yan et al., 2007; Gudendorf and Segers, 2012) to construct multivariate distribution functions. Some examples of this include Renard and Lang (2007), Schoelzel et al. (2008) and Durante and Salvadori (2010). A series of discussion papers about the use of copulas, their benefits and shortcomings, has been stimulated by Mikosch (2006). One more example of a method for the analysis of compound extreme events is using a Bayesian hierarchical framework to construct joint probability density functions of temperature and precipitation change (Tebaldi and Sansó, 2009).

For planning climate change related adaptation or mitigation measures with respect to the occurrence of compound extreme events, knowledge about their climate change signal is important. This information can be obtained by using climate models which can generate scenarios of the future evolution of the climate. As these are inherent to a number of uncertainties (e.g. future emissions or model uncertainty), multiple climate simulations, so called ensembles, are used (Collins, 2007; IPCC, 2010), which ideally cover the bandwidth of possible evolutions of the climate system. Through the use of ensembles, an important additional information about the change signal can be assessed, namely the uncertainty or probability of these changes. A further advantage of using ensembles is the generation of a broader statistical data basis which is especially important when looking at rare events, such as compound extremes.

Extreme events usually take place on different spatial scales. Some, like summer heat waves, depend mostly on large scale atmospheric circulations. Others, like convection based heavy precipitation events, are additionally influenced by regional scale climate forcings such as orography or soil-moisture interactions (Giorgi, 2006). In both cases, the magnitude of climate change usually depends strongly on the region. Global climate model ensembles, as for example the Coupled Model Intercomparison Project Phase 5 (CMIP5) ensemble (Taylor et al., 2012) used in the latest assessment report of the IPCC (IPCC, 2013) have a typical resolution of $\approx$ 100-300 and are not able to represent relevant small scale climate forcings and processes. Especially in regions with complex topography, which greatly affects the local climate, the coarsely resolved global climate data therefore needs to be translated to the regional scale. This can be done by using regional climate models which simulate a limited area but at higher resolution and with a much more detailed representation of regional climate forcings such as orography, soil types and land-use. This is an additional computational effort, but different studies (e.g., Feldmann et al., 2008; Feser et al., 2011) have shown that downscaling global climate model data with regional models can yield better results in comparison to observations depending on the quantity (e.g. for precipitation). This is also referred to as added value. For Europe, joint efforts such as the Prudence (Prediction of Regional Scenarios and Uncertainties for Defining European Climate Change Risks and Effects, Christensen et al., 2007) and ENSEMBLES (Van der Linden and Mitchell, 2009) projects have led to ensembles of regional climate simulations at a resolution of $\approx$50 km. Within the framework of the Coordinated Regional Climate Downscaling Experiment (CORDEX, Giorgi et al., 2009), an ensemble of even higher resolutions of $\approx$ 12 has been constructed. These projects have led to a better understanding of the European climate and the expected change in mean values but also in extremes. Compound extreme events, however, are largely underrepresented in the analyses. This thesis aims to advance the knowledge in this field and provide high resolution information about possible near future changes of compound extreme events. For this goal, a 12- member ensemble of climate simulations at an even higher resolution than the above mentioned projects of $\approx$ 7 km was generated covering central Europe. This permits an even better representation of the local conditions which influence the climate, especially for mountainous regions like the Alps or the Black Forest which are included in the model domain. The simulations cover two time periods, a reference period (1971-2000) for which the simulations can be compared to observations for validation and a time period in the near future (2021-2050), which is important for planning purposes of mitigation and adaptation strategies. The climate change signal can be calculated between these two time periods. Compound events are defined following the first of the above given definitions :"two or more extreme events occurring simultaneously" (IPCC, 2012) and the focus is on temperature and precipitation extremes, namely hot and dry extremes in summer and cold and wet extremes in winter. These extremes were chosen due to the availability of high resolution gridded observational data for the two variables for validation purposes. The main objective of this work is to answer 4 questions. The first three are:

1. Can the ensemble of regional climate simulations correctly simulate the statistical occurrence of compound extreme events for a reference time period (1971-2000)?

2. How will the statistical occurrence of compound extremes change between this reference period (1971-2000) and the near future (2021-2050), and how robust are the predicted changes? Can regions be identified which are especially susceptible to the change of extreme events?

3. Is there any added value from regional climate simulations at 0.0625 ° ($\approx$ 7 km) resolution in comparison to regional climate simulations at 0.44 ° ($\approx$ 50 km) resolution for the description of compound extreme events?

To answer these questions, three different methods, which investigate different aspects of compound extreme events, are applied to daily data of the high resolution climate ensemble. To address the question of added value, the same analysis is applied to a coarser resolved regional climate ensemble of 50 km resolution for the reference period.
The first two methods follow some of the ideas introduced above and describe compound extremes as concurrent threshold exceedances. The first method (further also referred to as type 1 extremes) defines compound extreme events by a joint absolute threshold exceedance of maximum/minimum temperature and precipitation. A fixed threshold for each variable and the respective extreme is defined which is the same for the reference period and the near future. One sector which is impacted by type 1 extremes (hot and dry extremes in summer) is the agricultural sector. Although the positive temperature trend might even be of advantage to the crop yield because it leads to an increase of the growing season length and overall warmer temperatures, the increase of absolute compound hot and dry extremes leads to a more erratic climate which can have a negative effect on the crop yield (Lavalle et al., 2009).

If compound extremes occur at critical stages of the growth cycle, they can cause large damage to the harvest. An example of a sector which is affected by cold and wet extremes in winter is the infrastructure sector. Cold and wet extremes in winter, e.g. heavy snowfall or heavy rain and subsequent freezing can be a threat to traffic, power supply and buildings among others.

The second method (further referred to as type 2 extremes) focuses on relative compound extreme events. Indices are calculated which describe daily standardized anomalies connected to temperature and precipitation, relative to the climatological mean state in the respective time period. A compound extreme event is again defined by a concurrent threshold exceedance of both indices. These extremes are not necessarily record breaking events but are defined as extreme deviations from the climatological mean state and as such are a measure for the variability of the climate system. Relative extremes mostly affect the ecosystem. Species usually have a "climatic envelope" which describes physiological thresholds of temperature and precipitation tolerance (e.g. Walther et al., 2002). If these thresholds are exceeded often, this can lead to a decrease or migration of certain species or even their extinction. Areas where this plays an important role are mountainous regions where species are often strongly adapted to the local climatic conditions and are highly affected by these change (Thuiller et al., 2005). The variability is also of importance since different studies imply that changes in variability play an important role for the frequency of extreme events (Katz and Brown, 1992; Schär et al., 2004).

The third method (further referred to as type 3 extremes), which was developed within this work, addresses an aspect of compound extreme events which has been largely neglected so far. It focuses on the dynamical behavior or temporal succession, i.e. not only on how many events occur but on how they are ordered. If, for example, two different regions show the same number of compound extreme hot and dry days in summer for a 30 year time period, the impact depends highly on their temporal succession, i.e., if there are a few days every summer or if most of the compound extreme events occur consecutively in one summer. Regarding climate change, this allows to investigate, whether or not a shift to a new climatological mean state also changes the dynamical behavior of compound extremes, a property which is up to now unknown and not intuitively accessible. Furthermore, differences between the predictability of compound extreme events can be assessed that relate to the chaotic behavior of the climate system. Information about changes in the dynamical behavior can be of importance to all sectors. Changes in these properties imply that even if an adaptation to a new "normal" climate state (with changed mean and variability) is possible (e.g. crops which need less water and can cope with higher temperatures or species which migrate to other regions), the respective sectors or species will still be subjected to a different frequency and duration of compound extreme events relative to the new "normal" state.

A synopsis of the results of the three methods for compound extremes allows to identify regions which are possibly more susceptible to changes of compound extreme events in the future and this knowledge can then be used for the planning of adaptation measures in these regions.

The occurrence of extreme events is largely triggered by certain atmospheric circulation patterns (e.g., Jacobeit et al., 2009). For a better understanding of these dependencies, weather types can be classified and their interrelation with extreme events studied (Kapsch et al., 2012; Riediger and Gratzki, 2014). Therefore, the objective weather type classification scheme developed by the German Weather Service (Bissolli and Dittmann, 2001) is applied to the ensemble of regional climate simulations. In a second step,the occurrence of weather patterns as well as their change signal are analyzed and set into relation to compound extreme events. This leads to the fourth question:

4. How are the compound extreme events dependent on different weather patterns and how will these change in the near future?

The thesis is divided into 10 chapters. Chapter 2 gives a general introduction to regional climate modeling, the regional climate model COSMO-CLM

(Consortium for Small Scale Modeling in CLimate Mode) used in this work and some information about ensembles. In Chapter 3, the regional climate ensemble generated in this work is introduced. Before analyzing compound extreme events, Chapters 4 - 6 give an overview of the mean values of the regional climate ensemble. Chapter 4 summarizes the performance of the ensemble with respect to temperature and precipitation imn comparison to observations for a time period in the past, 1971-2000. Due to biases in both temperature and precipitation time series, the data is bias corrected. The bias correction methods and their effect on the climate change signal are analyzed in Chapter 5 and in Chapter 6 the change signal of mean values between this past time period (1971-2000) and the near future (2021-2050) are presented. Chapters 7 - 9 focus on the analysis of compound extreme temperature and precipitation events with the above introduced methods, Chapter 7 focuses on absolute threshold exceedances (type 1 extremes) and Chapter 8 on relative threshold exceedances (type 2 extremes) of temperature and precipitation indices. A special focus lies on the newly developed method focusing on the dynamical behavior of compound extreme events (type 3 extremes) presented in Chapter 9. An analysis of influencing factors in the form of weather patterns is given in Chapter 10. All Chapters include a short summary of the main findings, a more thorough summary and discussion of the results, especially of Chapters 7 - 10 can be found in Chapter 11.

2 Background

The following chapter includes the theoretical background of the generation of an ensemble of regional climate simulations with the COSMO-CLM regional climate model. In Section 2.1, a short introduction to regional climate modeling is given followed by an overview of the COSMO-CLM model in Section 2.2. Basic assumptions and statistical methods used for the generation and evaluation of the ensemble are explained in Section 2.3.

2.1 Regional climate modeling

Climate affects most human and natural systems and therefore information about the present state and possible changes in climatic variables are of great interest to society. Impacts of climate change are usually region specific. This calls for high resolution information about possible change scenarios. Although the resolution of global climate models (further referred to as GCMs) has been rapidly increasing (Mizielinski et al., 2014), most state of the art GCMs are still run at resolutions in the order of hundreds of kilometers (Meehl et al., 2007; Taylor et al., 2012). At this resolution the orography is smoothed and interactions of topography and land surface, which affect the local scale climate, are not included. Especially in regions with complex topography high resolution data is therefore important (Giorgi, 2006). The resolution furthermore plays a great role when considering extreme events, which are smoothened out over a greater area in the coarser resolved models (Giorgi, 2006). Generally, there are two different ways of downscaling global climate model data to the regional scale and, thus, increasing the resolution of the coarse GCM output. One is statistical downscaling where relations between large-scale climate variables and regional local features are derived from observational datasets and then used to statistically estimate local variables from their global counterparts (for an overview see e.g. Wilby et al., 1998). The main advantage of the statistical methods is that they are computationally inexpensive and that a potential bias in the results can be corrected. The availability of good quality observational data can however pose a problem and limit the variables that can be downscaled. Furthermore, the statistically downscaled variables are not necessarily a physically consistent set of variables and it is assumed that the large-scale - regional-scale relations are stationary over time. The second way to downscale data from global climate models and the one used in this work, is dynamical downscaling by using a regional climate model (RCM). This yields a set of variables which is physically consistent in time and space.

Regional climate models only compute the climate for the region of interest but with a higher spatial (and usually also temporal) resolution than global climate models. Since the regional processes are not independent of global circulation patterns, data from global climate models or reanalysis products is regularly fed to the regional climate model at the lateral domain boundaries (also called boundary data) and for initialization. The boundary data also contains the response of the global circulation patterns to large scale radiative forcing prescribed by the emission scenarios (Nakicenovic and Swart, 2000; Moss et al., 2010). One limitation that regional climate models have is that they can only be as good as the driving global climate model. This has been phrased as "garbage in, garbage out" (Giorgi and Mearns, 1999). If for instance, the driving global climate model cannot capture some important processes and circulation patterns such as the North Atlantic Oscillation (NAO) which greatly affects the climate in Europe or global trends of atmospheric variables are not represented well, the regional climate model will not be able to capture these processes and their effects, either.

First attempts at regional downscaling of climate data, using limited area models, were conducted in the late 80s and early 90s of the last century (Dickinson et al., 1989; Giorgi, 1990; Jones et al., 1995). With increasing availability of high performance computing, the time periods and regions simulated became longer and larger. Today there is a vibrant regional climate modeling community and different international regional downscaling projects including multiple regional climate models (e.g. ENSEMBLES, Van der Linden and Mitchell (2009); CORDEX, Giorgi et al. (2009); NARCCAP, Mearns et al. (2012)). A more thorough introduction to regional climate modeling can be found in McGregor (1997), Wang et al. (2004), Giorgi (2006) and Laprise (2008).

Climate projections are usually only statistically evaluated over longer time periods because the global climate data used as initial and boundary data is generally not synchronized with observations (exceptions are reanalysis products where a large number of observations have been processed by a model to output a consistent set of variables) . This is due to the initialization process of the GCM. In the first stage, the GCM runs with constant external forcing (corresponding to the value of preindustrial emissions) for a few hundred years until all the components of the model (atmosphere, ocean, sea ice, etc.) are approximately in an equilibrium state. In a second step, one day of this run is taken as initial condition for a historical run externally forced by observed 20th century emissions. The date of this run is set according to the external forcing and this usually starts in 1850. For climate projections or climate scenarios, a day of one of the last years of the historical model run (e.g. from 2006 for CMIP5 Taylor et al., 2012) is used as initial state and emission scenarios describing the possible future developments of emissions (Nakicenovic and Swart, 2000; Moss et al., 2010) are used as radiative forcing (see also e.g. http://climate4impact.eu). Even though modelers can try to pick initial states that fit well with the observed climate for the historical and scenario runs, it will most likely not be the same as the actual state of the climate system on that day. Therefore, it is not possible to evaluate climate projections for specific days. Instead, they are evaluated over longer time periods where variations on yearly timescales are filtered out but long term climatic trends, which are induced by the changes in external forcing, can still be identified. Following the recommendations by the World Meteorological Organization (WMO), time periods of 30 years are generally considered for the analysis (Trewin et al., 2007). Therefore in regional climate modeling often two (or more) 30 years time periods are chosen; one within the time period of the historical GCM run and one (or more) for the time period of the scenario run. A climate change signal between the different time periods can then be analyzed.

2.2 The COSMO-CLM regional climate model

COSMO-CLM (Consortium for Small Scale Modeling in CLimate Mode) is the climate version of the numerical weather prediction model COSMO which has its origins in the Lokal Modell (LM) of the German Weather Service DWD.
The COSMO-LM community provides a thorough documentation of the model system (Doms et al., 2011b,a). The following paragraphs give a short summary of this model description. In Section 2.2.1, the coordinate system of the COSMO model is introduced followed by a description of the thermo-hydrodynamic equations which form the core of the model (Section 2.2.2) and the main parametrization used in the model (Section 2.2.3). The numerical solution is explained in Section 2.2.4 and Section 2.2.5 contains an introduction to the initial and boundary conditions used to run the model.

2.2.1 Coordinate system

The COSMO-CLM uses spherical coordinates because the curvature of the earth can usually not be neglected for the typical domain sizes of several thousand square kilometers. To avoid singularities due to convergence of meridians at the poles, a rotated coordinate system, with the geographical longitude λ and the geographical latitude φ, is used, where the pole is rotated outside of the model area. The crossing point of the equator and the zero meridian is ideally positioned in the center of the model domain as this minimizes the grid distortion. Furthermore, a terrain following vertical coordinate ξ is used. This highly simplifies the description of the lower boundary because it corresponds to the earth's surface. The upper boundary is an even surface.

2.2.2 Fundamental model equations

- horizontal wind velocity (u, v)

$$\begin{aligned}\frac{\partial u}{\partial t} &= -\left\{\frac{1}{acos\varphi}\frac{\partial E_h}{\partial \lambda} - vV_a\right\} - \dot{\zeta}\frac{\partial u}{\partial \zeta} \\ &\quad - \frac{1}{\rho acos\varphi}\left(\frac{\partial p'}{\partial \lambda} - \frac{1}{\sqrt{\gamma}}\frac{\partial p_0}{\partial \lambda}\frac{\partial p'}{\partial \zeta}\right) + M_u\end{aligned} \qquad [2.1]$$

$$\frac{\partial v}{\partial t} = -\left\{\frac{1}{a}\frac{\partial E_h}{\partial \varphi} - uV_a\right\} - \dot{\zeta}\frac{\partial v}{\partial \zeta} - \frac{1}{\rho a}\left(\frac{\partial p'}{\partial \varphi} - \frac{1}{\sqrt{\gamma}}\frac{\partial p_0}{\partial \varphi}\frac{\partial p'}{\partial \zeta}\right) + M_v \qquad [2.2]$$

- vertical wind velocity (w)

$$\begin{aligned}\frac{\partial w}{\partial t} &= -\left\{\frac{1}{acos\varphi}\left(u\frac{\partial w}{\partial \lambda} + vcos\varphi\frac{\partial w}{\partial \varphi}\right)\right\} - \dot{\zeta}\frac{\partial w}{\partial \zeta} + \frac{g}{\sqrt{\gamma}}\frac{\rho_0}{\rho}\frac{\partial p'}{\partial \zeta} \\ &\quad + M_w + g\frac{\rho_0}{\rho}\left\{\frac{(T - T_0)}{T} - \frac{T_0 p'}{T p_0} + \left(\frac{R_v}{R_d} - 1\right)q^v - q^l - q^f\right\}\end{aligned} \qquad [2.3]$$

- perturbation pressure ($p' = p - p_0(z)$)

$$\frac{\partial p'}{\partial t} = -\left\{\frac{1}{acos\varphi}\left(u\frac{\partial p'}{\partial \lambda} + vcos\varphi\frac{\partial p'}{\partial \varphi}\right)\right\} - \dot{\zeta}\frac{\partial p'}{\partial \zeta} + g\rho_0 w - \frac{c_{pd}}{c_{vd}} pD \quad [2.4]$$

- temperature (T)

$$\frac{\partial T}{\partial t} = -\left\{\frac{1}{acos\varphi}\left(u\frac{\partial T}{\partial \lambda} + vcos\varphi\frac{\partial T}{\partial \varphi}\right)\right\} - \dot{\zeta}\frac{\partial T}{\partial \zeta} - \frac{1}{\rho c_{vd}} pD + Q_T \quad [2.5]$$

- water vapor (q^v)

$$\frac{\partial q^v}{\partial t} = -\left\{\frac{1}{acos\varphi}\left(u\frac{\partial q^v}{\partial \lambda} + vcos\varphi\frac{\partial q^v}{\partial \varphi}\right)\right\} - \dot{\zeta}\frac{\partial q^v}{\partial \zeta} - \left(S^l + S^f\right) + M_{q^v} \quad [2.6]$$

- liquid and solid forms of water ($q^{l,f}$,l:liquid,f:solid)

$$\frac{\partial q^{l,f}}{\partial t} = -\left\{\frac{1}{acos\varphi}\left(u\frac{\partial q^{l,f}}{\partial \lambda} + vcos\varphi\frac{\partial q^{l,f}}{\partial \varphi}\right)\right\} - \dot{\zeta}\frac{\partial q^{l,f}}{\partial \zeta} - \frac{g}{\sqrt{\gamma}}\frac{\rho_0}{\rho}\frac{\partial P_{l,f}}{\partial \zeta} + S^{l,f} + M_{q^{l,f}} \quad [2.7]$$

- total density of air (ρ)

$$\rho = p\left\{R_d\left(1 + \left(\frac{R_v}{R_d} - 1\right)q^v - q^l - q^f\right)T\right\}^{-1} \quad [2.8]$$

with the following variables:

$\sqrt{\gamma} = \frac{\partial p_0}{\partial \zeta}$	variation of reference pressure with ζ
$E_h = \frac{1}{2}(u^2 + v^2)$	kinetic energy of horizontal motion
$V_a = \frac{1}{acos\varphi}\left\{\frac{v}{\lambda} - \frac{\partial}{\partial \phi}(ucos\phi)\right\} + f$	absolute vorticity
q^v, q^l, q^f	specific water vapor/ liquid water/ ice content
T_o, p_o, ρ_o	base state values of temperature, pressure and density
f	Coriolis parameter
a	earth radius
ρ	density
ζ	terrain following vertical coordinate
$\dot{\zeta}$	contravariant vertical velocity
φ, λ	rotated latitude, longitude
$M_u, M_v, M_w, M_{q^v}, M_{q^l}, M_{q^f}$	sub-grid scale (turbulent) exchange processes
D	divergence of the wind field
Q_t	diabaitc heating term
c_{vd}, c_{pd}	specific heat at constant pressure of water vapor, dry air
P^l, P^f	precipitation fluxes
S^l, S^f	cloud microphysical sources and sinks due to phase change
R_v, R_d	gas constants for dry air and water vapor

The core of the COSMO-CLM model is a set of primitive thermo - hydrodynamic prognostic equations which are based on the conservation of momentum, total mass and heat. The derivation of the final set of model equations (eq. 2.1 - 2.8) is very well documented in Doms et al. (2011a). Vertical and horizontal wind components (u, v), temperature (T), pressure (p), water vapor(q^v), liquid and solid water (q^l, q^f) are calculated as prognostic variables. Additionally the total density of air (ρ) is calculated by a diagnostic equation.

2.2.3 Parametrizations and lower boundary conditions

Atmospheric processes take place on different temporal and spatial scales. Some, like large scale atmospheric circulations, are orders of magnitude larger than the grid spacing and can be explicitly calculated. Others, e.g. convection, however, are often sub-grid scale processes and cannot be explicitly resolved by the model. These processes are described by parametrizations depending on variables calculated by the model. The lower boundary conditions are obtained from a land surface model. All parametrizations used in the COSMO-CLM are described in detail in Doms et al. (2011b). In the following a short summary of the most important of these parametrizations is given.

Sub-grid scale turbulence:
The sub-grid scale turbulence is parametrized by using a closure at level 2.5 (Mellor and Yamada, 1982) on the prognostic equation for the turbulent kinetic energy (TKE) . Alternatively a 2.0 order closure can be applied. The sub-grid scale turbulence is included in the prognostic equations 2.1 - 2.3 by the sub-grid scale (turbulent) exchange processes M_u, M_v and M_w.

Turbulent flow near the surface:
Turbulent fluxes allow exchange processes of momentum, heat and humidity between the atmosphere and the earth's surface. The parametrization is related to the sub-grid scale turbulence scheme. The surface layer, which is defined as the layer of air between the earth's surface and the lowest model level, is divided into three parts, a laminar-turbulent sublayer, the roughness layer and a Prandtl layer (Heise, 2002).

Clouds and precipitation:

The two-category ice scheme used in the COSMO-CLM differentiates four hydrometeor categories in addition to water vapor: cloud water, cloud ice, rain and snow. The possibility of including graupel is given by an optional three category ice scheme. The transport of precipitation is three dimensional and microphysical processes are accounted for. Sub-grid scale clouds are calculated by an empirical function depending on humidity and height (Heise, 2002).

Parametrization of moist convection:

There are three different convection schemes implemented in the COSMO-CLM: the Tiedtke scheme (Tiedtke, 1993), the Kain-Fritsch scheme (Kain, 1993) and the Bechtold scheme (Bechtold et al., 2001). They differ in the closures they use as well as the triggering criteria for convection and the processes influencing detrainment and entrainment (Smodydzin, 2004). The Tiedtke scheme is the standard scheme used.

Radiative processes:

The radiation scheme solves the δ two-stream version of the radiative transfer equation. The effects of scattering, absorption, and emission by cloud droplets, aerosols, and gases in each part of the spectrum are incorporated in the parametrization (Ritter and Geleyn, 1992). The scheme uses eight spectral intervals (Heise, 2002).

Parametrization of Sub-Grid Scale Orography:

External data such as orography, soil type, vegetation cover and land-sea mask are available from external data sets and are preprocessed to the desired resolution and rotated grid as described in Smiatek et al. (2008). The orography is taken from the GLOBE (The Global Land One-km Base Elevation Project) and the land classes from the GLC2000 dataset (Bartholome and Belward, 2005).

The Soil and Vegetation Model TERRA:

The lower boundary conditions are obtained from the soil vegetation model TERRA (Schrodin and Heise, 2002). It uses prognostic equations for soil temperature and soil water content and in climate mode is usually run with ten vertical soil layers.

2.2.4 Numerical solution

To solve equations 2.1 - 2.8, a numerical solution is necessary. For this, the prognostic equations are discretized in space and time using a finite difference method. A horizontally equidistant, three dimensional model Arakawa-C/Lorenz grid is used in the COSMO model (see Fig. 2.1). A grid point (i,j,k) is defined as the center of a grid cell with lengths $\delta\lambda$,$\delta\varphi$ and $\delta\zeta$. Scalar model variables, such as temperature T or pressure p, are defined on the grid points, the wind component vectors are defined on the edges of the grid points ($\lambda_{i\pm1/2}$,$\phi_{i\pm1/2}$,$\zeta_{i\pm1/2}$). For the time discretization, a third order Runge-Kutta time-stepping scheme (Wicker and Skamarock, 2002) is used.

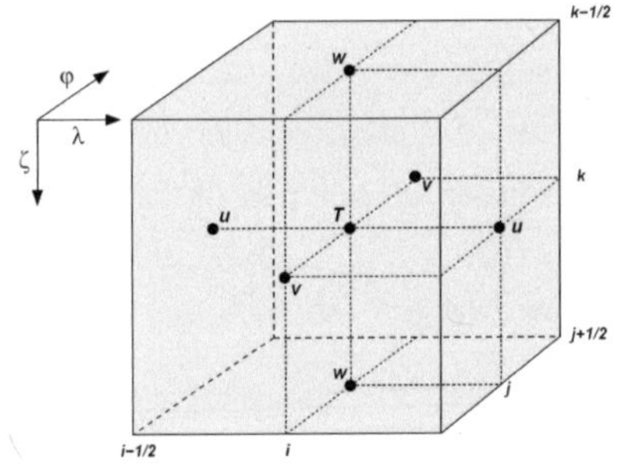

Figure 2.1: Exemplary grid box of the COSMO-CLM with Arakawa-C/Lorenz staggering (from Doms et al., 2011a, p. 52)

2.2.5 Initial and boundary conditions

A regional climate model has lateral and horizontal boundaries. At the start of and during a simulation, the conditions of the atmosphere and soil at these boundaries is needed. The atmospheric prognostic variables are obtained by interpolating data from a coarse grid climate model or reanalysis to the desired resolution. This is done by using the preprocessor program INT2LM (Schättler and Blahak, 2015). At the lateral boundaries, a one-way interactive nesting is used. The time evolution of the prognostic variables is taken from the preprocessed global dataset. The difference in resolution causes numerical problems and can lead to numerical noise at the boundaries. To prevent a propagation of the noise into the center of the model domain, a relaxation scheme by Davies (Davies, 1976, 1983) is used. A sponge zone is introduced where the model variables are modified until they blend in with the variables of the forcing dataset. This relaxation scheme is applied to all prognostic variables except for the vertical wind velocity for which a free slip lateral boundary condition is specified. The sponge zone usually consists of ten grid points at all lateral boundaries. There is no mass transfer across the top boundary, the surface is treated as rigid. To suppress wave reflections, a Rayleigh damping scheme may be applied. The lower boundary is described by the soil-atmosphere-vegetation model TERRA.

2.3 Ensembles

Ensembles are a set of comparable model simulations. In the field of climate science, ensembles are an important tool to quantify the uncertainties of climate simulations and for allowing to decide how robust a given outcome is (e.g. Collins, 2007; IPCC, 2010). The climate system is a very complex system including many non-linear effects on different spatial and temporal scales. Although atmospheric dynamics can be described by a few known differential equations, models will always be approximations of the reality and inherently include some uncertainty. External parameters which go into the simulations are a further source of uncertainty. One single climate model run is therefore just one possible evolution of the climate under certain assumptions and not a reliable source for climate information. Ensembles of simulations can be constructed to include the range of uncertainties. In the following, some of the uncertainties of climate simulations are listed and mentioned how they can be addressed. The link to regional climate model ensembles is given subsequently.

1. Structural uncertainty: Climate models are always a simplification of the real climate system. To keep the computational effort within reasonable limits, not all processes can be included and each modeling center needs to decide where to make amends, for example on how complex the ocean model should be. Furthermore, there are still mechanisms which are not fully understood and are implemented differently in different models.
   →How to address this uncertainty: multiple models
   By using models with different structural components, a so called multi-model ensemble (Tebaldi and Knutti, 2007), the structural uncertainty can be addressed. This uncertainty can be reduced to some extent with better knowledge of the climate system and the availability of higher computational power. This would allow a more sophisticated description of certain processes or better resolution of the model components such as the ocean model or a land model.

2. Parameter uncertainty: Processes taking place on time and/or spatial scales smaller than the model resolution cannot be explicitly resolved by the model. Therefore parametrizations exist which describe these processes in terms of the resolved model variables. Often, there is more than one parametrization available for a certain process.
   →How to address this uncertainty: perturbed physics

To address the uncertainty of the parametrization, different parameters can be perturbed or whole parametrization schemes changed.

3. Uncertainty in boundary conditions: The boundary condition which causes the most uncertainty is the radiative forcing which takes into account the external variability induced by anthropogenic forcing. For time periods that lie in the future, this radiative forcing can only be estimated. Several emission scenarios or representative concentration pathways (RCP) corresponding to different possible future changes of social and economic development are put together by a group of experts which are supposed to span a range of possible future radiative forcing.
   →How to address this uncertainty: different emission scenarios
   To address the uncertainty range due to emission scenarios, different emission scenarios can be used as external forcing data. This uncertainty is hard to reduce since human development is especially hard to predict.

4. Uncertainty in initial conditions: The climate system underlies a natural variability, natural fluctuations which originate in the nonlinear interactions between different parts of the climate system. These are quasi periodically occurring events which take place on different time scales. One example of fluctuations on shorter time scales are the El Niño-Southern Oscillation (ENSO, Allan et al., 1996) or the North Atlantic Oscillation (NAO, Hurrell et al., 2001) which take place on yearly to decadal time scales.
   →How to address this uncertainty: different realizations
   The internal variability can be addressed by using the same model with fixed parameters and external data but using different states of the GCM as initial data, thus varying the phase of the internal variability (see Section 2.1). Each initialization is then called a different realization of the same climate simulation. The uncertainty due to

> the internal variability cannot be reduced. Deser et al. (2012) have "estimated this internal variability to account for at least half of the inter-model spread in projected climate trends during 2005-2060 in the CMIP3 multi-model ensemble".

A regional climate ensemble is generally constructed by downscaling data from a global climate model or ensemble. The first two uncertainties can also be addressed with a regional model while the last two uncertainties of the above list cannot be explicitly addressed by the regional climate model but by downscaling the respective ensemble of global climate simulations. A review of the uncertainties in regional climate modeling is given in Feser et al. (2011). Using multiple regional climate models to downscale a global climate model usually requires collaboration between different climate modeling groups because of the high computational demand. Even then, the whole matrix of possible GCM-RCM combinations is usually never fully sampled (e.g. ENSEMBLES (Van der Linden and Mitchell, 2009) or NARCCAP (Mearns et al., 2012) projects). A further ensemble generation method for regional climate models which was used in this work is the Atmospheric Forcing Shifting (AFS), introduced by Sasse and Schädler (2014). In this case, the GCM data interpolated to the RCM model grid is shifted in each cardinal direction by two grid points and then used as forcing data, thus changing its location with respect to orographic features. This method accounts for the uncertainty in positioning of synoptic system when downscaling and mostly affects precipitation.
Most climate ensembles today (e.g. the CMIP global ensembles (Meehl et al., 2007; Taylor et al., 2012) or the ENSEMBLES (Van der Linden and Mitchell, 2009) and CORDEX (Giorgi et al., 2009) regional climate ensembles) are so called "ensembles of opportunity" (Tebaldi and Knutti, 2007; Annan and Hargreaves, 2010) and include a mixture of the above mentioned possible ensemble generation techniques "but are not designed to sample uncertainties in a a systematic way" (IPCC, 2010) and most likely do not include the

full uncertainty range. The reason for this is the high computational effort associated with climate modeling, leading to ensembles that consist of a collection of available simulations. It should be noted that there is always the possibility of changes outside of the corridors of the ensemble spread.

Besides quantification of uncertainties, a further reason to use an ensemble of climate simulations is the broadening of the statistical data basis. This is especially important when looking at rare events such as extremes and compound extreme events. In these cases, considering only one model might not be sufficient for a statistical analysis of the results.

Finding the optimal size and composition of climate ensembles is an ongoing research topic and also depends on the assumptions made for the underlying framework (e.g., Evans et al., 2013; Haughton et al., 2014). Besides the availability of data, especially on regional scales at high resolution, the question of ensemble size versus statistics and variables considered are not settled satisfactorily (Ferro et al., 2012).

The statistical evaluation of an ensemble can have different underlying frameworks. Two common ones are the following:

1. "truth plus error paradigm" (IPCC, 2010): In this case, each ensemble member is assumed to be sampled from a distribution centered around the truth. This implies that, as more models are included in the ensemble, the error should be reduced and the ensemble mean approaches the value of the observations.

2. "indistinguishable paradigm" (Annan and Hargreaves, 2010): Here ensemble members and observations are all seen as exchangeable and the ensemble mean converges to the statistical center of the distribution of all ensemble members which is generally not equal to the observations. This approach allows a more probabilistic interpretation of ensembles.

Both approaches assume independent model members. There are different types of independence, independence of the model input, structural independence and model output independence (Bishop and Abramowitz, 2013).

As measure for input and output independence, correlation is mostly used. Generally, state of the art ensembles never satisfy all three independence criteria. Bishop and Abramowitz (2013) have analyzed the CMP3 ensemble for independence and found it to be not fully independent. Some of these dependencies are hard to eliminate though, because some models share the same physics through parametrization, some even whole components. For the analysis, independence is usually assumed.
Another ongoing research topic is the question of whether or not to assign weights to models when combining the members of an ensemble. A general assumption that needs to be made when weighting models is that the performance of the models is persistent through time, meaning that a good model in the verification time period is also a good model in the future. In order to meet this assumption, the weights have to be deduced with care as not to give a model a high weight for the wrong reasons (e.g. if errors of the GCM cancel out by errors of the RCM). Additionally, a model performing well for one variable might not be good for other variables. There are different approaches for deducing performance-based model weights. Among them are the bias with respect to observations and weights depending on the models ability to capture certain processes or trends. Within the ensembles projects, a weighting scheme was developed which combines several performance metrics (Van der Linden and Mitchell, 2009; Kjellström et al., 2010). There are however also critics of model weighting (e.g. Weigel et al., 2010; Déqué and Somot, 2010) and Bukovsky et al. (2013) found that using metrics as in the ENSEMBLES project does not yield a clear improvement compared to the unweighted ensemble mean. If a "family" of model runs is included in an ensemble (e.g. different runs with slightly perturbed parameters), these are often weighted in the ensemble in order not to violate the assumption of model independence.

Concluding this section about ensembles, the ensemble measures used throughout this work are listed:

- Ensemble mean: average of the different ensemble members for a given property. Whenever statistical parameters (e.g. indices) are calculated, they are calculated for each ensemble member individually before the ensemble mean is derived. Otherwise the internal variability would cancel out. A number of studies have shown that the ensemble mean of a multi-model ensemble is often closer to observations than individual models (Gleckler et al., 2008; Weigel et al., 2008; Reichler and Kim, 2008).
- Ensemble spread: The ensemble spread is a measure of the uncertainty range that is spanned by the different ensemble members. Measures for the ensemble spread are
  - standard deviation $\sigma = \sqrt{\sum (x - \bar{x})^2 / n - 1}$
  - interquartile range: this is the difference between the 75th quantile and the 25th quantile of the distribution over the ensemble values. It is a robust measure for the ensemble spread.
- Ensemble consistency: Measure of the uncertainty of the climate change signal, which is important information for planning purposes. In this work, the consistency of a climate change signal is calculated as proposed by Feldmann et al. (2012), by subtracting the number of ensemble members with a negative signal greater than a certain threshold from the number of ensemble members with a positive signal (greater than the threshold) and normalizing it by the total number of ensemble members. An ensemble consistency of 100% /-100% therefore signifies that all members show a positive/negative change, whereas a consistency of 0% implies that the models are discordant or that the change is smaller than the threshold for all ensemble members. As threshold, a relative change signal of 10% is used. Throughout

this work, a change signal is referred to as robust or consistent if the ensemble consistency is at least 50%, which means that if the change signals are all higher than the threshold value (10%), 75% of the ensemble members show the same sign of the change signal.

3 The COSMO-CLM ensemble

The ensemble analyzed in this work consists of 12 members. Four ensemble members come from an ensemble already existing COSMO-CLM ensemble at the Karlsruhe Institute of Technology. The eight additional members were generated using the regional climate model COSMO-CLM (described in Chapter 2.2), model version cosmo_090213_4.8_clm17. Three members were generated within this work and 5 members were partly taken from work by Sasse and Schädler (2014) and finalized within this work. A double nesting approach was used to downscale GCM data to a resolution of 0.0625° $\approx$ 7 km. This implies a first downscaling step of the global data to a lateral resolution of 0.44° $\approx$ 50 km for a domain covering Europe (108x110 grid points, show in Fig. 3.1) and then in a second step, using this 0.44° COSMO-CLM simulation as initial and boundary data for a simulation for a smaller

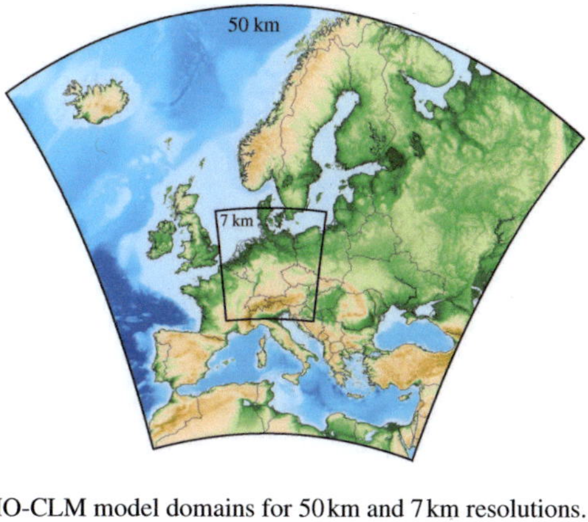

Figure 3.1: COSMO-CLM model domains for 50 km and 7 km resolutions. Image Source: Hans Schipper

domain covering only central Europe (165x200 grid points, see Fig. 3.1) with a higher resolution of 0.0625°. All simulations were performed on the Cray XE6 at the High Performance Computing Center Stuttgart (HLRS).
Data from 6 different GCMs (one with different realizations) was used for the simulations. Furthermore, the AFS method (see Section 2.3) was applied. An overview of the GCMs, their resolution and the emission scenario used are given in Table 3.1. The simulations using the climate models ECHAM5 (Fifth generation of the ECHAM general circulation model, acronym for

Table 3.1: Global climate models and reanalysis data used as initial and boundary data for the COSMO-CLM Ensemble.

Name	Atmospheric-model	Institute	Emission-scenario	resolution atmosphere	reference
ECHAM5	ECHAM5/MPIOM	Max Planck Institute for Meteorology, Germany	A1B	1.8°	Roeckner et al. (2003)
ECHAM6	MPI-ESM-LR	Max Planck Institute for Meteorology, Germany	RCP8.5	1.8°	Stevens et al. (2013)
CCCma3	CGCM3.1	Canadian Centre for Climate Modelling, Canada	A1B	1.8°	Scinocca et al. (2008)
EC-EARTH	EC-EARTH	EC-EARTH Consortium, Europe	RCP8.5	1.125°	Hazeleger et al. (2010)
CNRM-CM5	CNRM-CM5	National Centre for Meteorological Research, France	RCP8.5	1.4°	Voldoire et al. (2013)
Hadley	HadGEM2-ES	Met Office Hadley Centre, UK	RCP8.5	1.25° × 1.875°	Collins et al. (2011)
ERA40		ECMWF, UK	Reanalysis	1.125°	Uppala et al. (2005)

ECMWF and Hamburg) and CCCma3 (third generation atmospheric general circulation model of the Canadian Centre for Climate Modelling and Analysis) with emission scenario A1B and the ERA40 reanalysis as boundary data were all generated within the CEDIM project (Center for Disaster Management and Risk Reduction Technology, Berg et al., 2013; Wagner et al., 2013). All simulations with ECHAM6 (including the ones with AFS) were partly simulated within a project of the Helmholtz-Network REKLIM (Regional Climate Change, see Sasse et al., 2013) and partly within this work. All other simulations using global climate data with the emission scenario RCP8.5 were added to the ensemble as part of this work.

Two different time periods were simulated - one, covering the years 1971-2000, which is further called the reference time period and one, which is referred to as the near future covering the years 2021-2050. This time period is relevant for planning purposes. The reference period serves for evaluation of the model results and as a reference against which changes for the near future are calculated. The simulations were started three years before the respective time period in order to achieve an equilibrium between atmosphere and soil components.

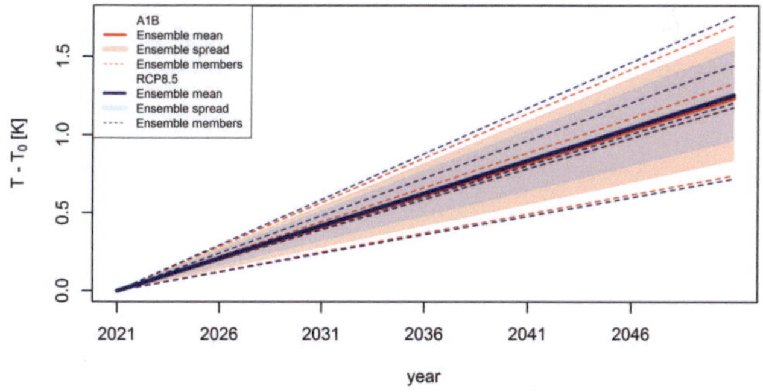

Figure 3.2: Mean linear temperature trends of ensemble members over the HYRAS domain (see Fig. 4.1) for the near future (2021-2050). Ensemble members are grouped according to the emission scenario underlying the GMC simulations used as boundary data.

This ensemble is clearly an "ensemble of opportunity" (see Section 2.3), as different ensemble generation techniques are combined. The boundary data used for the generation of the COSMO-CLM ensemble uses two different emission scenarios (see Table 3.1), A1B and RCP8.5. For the time periods considered in this work, the uncertainty of the emission scenario does not play a dominant role and the ensemble spread due to different driving GCMs is larger than the spread caused by the different emission scenarios (see Fig. 3.2). Therefore, simulations with both emission scenarios are combined.

For this work, no performance-based weighting measures were used. The ensemble members were, however, tested for independence of the model output. This was done by calculating pairwise correlations of the yearly time series between the different ensemble members for both temperature and precipitation. The resulting correlation matrices are pictured in Fig. 3.3. The model run using the GCM ECHAM6 and the AFS-shifted ECHAM6 model runs exhibit a high correlation for both variables. For all of this work, the ECHAM6 runs were therefore weighted as one, each receiving a weight of 1/5 whereas all other models were assumed to be independent and assigned a weight of 1, leading to an effective ensemble size of eight.

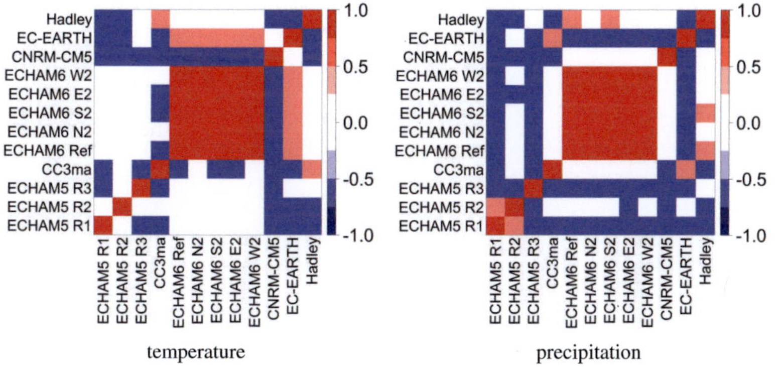

Figure 3.3: Correlation of mean yearly means over the HYRAS domain (see Fig. 4.1) between ensemble members in the reference period, 1971-2000. Left side: temperature, right side: precipitation.

4 Validation of ensemble mean values

This section is meant to give an overview of the performance of the ensemble regarding mean values of the ensemble with a resolution of 0.0625°. The ensemble is hereafter referred to as the 7 km ensemble. Only mean, minimum and maximum temperature and precipitation sums are evaluated since those are the variables that are used later for analyzing compound extreme events (in Chapters 7, 8 and 9). A validation of a larger set of model variables can be found in Sedlmeier and Schädler (2014) for the state of Baden-Württemberg. As explained in Section 2.1, 30 year time periods are used for the validation of the ensemble, means over these 30 year time periods are also referred to as climatological means/maxima/minima/sums. The ensemble spread refers to the ensemble mean plus/minus one standard deviation σ, unless

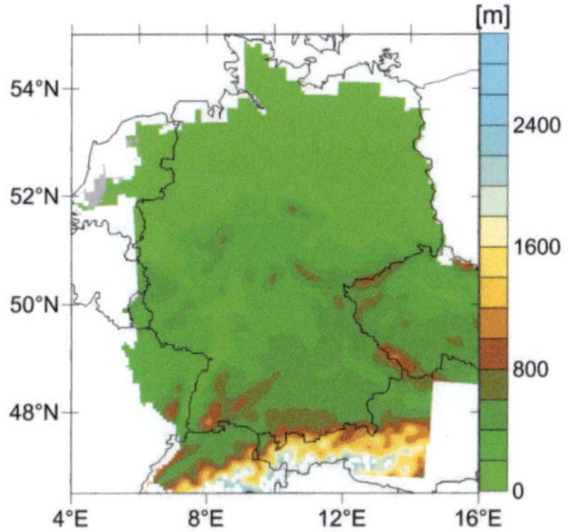

Figure 4.1: Investigation area, further referred to as "HYRAS domain", elevation [m].

stated otherwise. All validations are for gridpoints for which both E-Obs and HYRAS observational data exist, further referred to as “HYRAS domain” (see Fig. 4.1). The ensemble is validated for mean monthly means/sums as well as mean summer/winter sums. In this and the following two chapters, summer stands for the hydrological summer from May to October and winter for the hydrological winter from November to April.

4.1 Observational data for validation

4.1.1 HYRAS dataset

The HYRAS dataset (Rauthe et al., 2013) is a gridded daily dataset for mean daily temperature and daily precipitation sums covering the river catchements in Germany and the neighboring countries. The spatial resolution of the dataset is 1 km^2 and it is based on 6200 stations. It is available for the time period 1951-2006. The dataset was generated as part of the KLIWAS project (http://www.kliwas.de) by applying the REGNIE method (“REGionalisierung der NIEderschlagshöhen”, regionalized precipitation amount). This method consists of two steps. First the background climatological fields are calculated by multiple linear regression, taking geographical position, orientation and absolute value of wind exposure into account. In a second step, the quotients of the daily data and the background field are interpolated using inverse distance weighting and subsequently multiplied by the background field. The main advantage of this method is that the measured precipitation amounts are conserved (Rauthe et al., 2013). This is especially useful for extreme events which are found unchanged in the gridded field. The HYRAS dataset was bilinearly interpolated to the 7 km grid for validation of the COSMO-CLM ensemble.

4.1.2 E-Obs dataset

The E-Obs dataset (Haylock et al., 2008) is a gridded daily observation set for precipitation sums, minimum, maximum and mean air temperature. The dataset covers land points for Europe ($\approx$ 25N-75N, 40W-75E) for the time period 1950-2006. The dataset was generated as part of the ENSEMBLES project (http://www.ensembles-eu.org) and based on data of up to 2316 stations (depending on the time period). The interpolation of station data was done in a three stage process. First, the monthly mean precipitation and monthly mean temperatures were interpolated by using three dimensional thin-plate splines. In a second step, daily anomalies were interpolated (using the universal Kriging method for precipitation and Kriging with an external drift for temperature). Finally these monthly and daily interpolated values were combined. The uncertainty of this dataset depends strongly on the number of stations which went into the interpolation and therefore, on the region. The station density is highest in Switzerland, the Netherlands and Ireland and rather low in Spain and the Balkans which leads to an over-smoothing in these areas. This especially affects extremes of daily temporal resolution and has to be taken into account when using the E-Obs data for validation purposes. Version 10 of the dataset was used for all evaluations in this work. For the validation of the results, the E-Obs dataset was bilinearly interpolated to the 7 km grid. For temperature data, a height correction was applied when remapping to the 7 km grid by multiplying the elevation difference between the remapped E-Obs elevation and the 7 km COSMO-CLM grid with a constant lapse rate of 6.5 K/km for each grid point.

4.2 Mean temperature

The yearly cycle of the areal mean climatological monthly temperature means is depicted in Fig. 4.3 for the HYRAS domain. The ensemble mean can reproduce the yearly cycle of the observations fairly well but shows

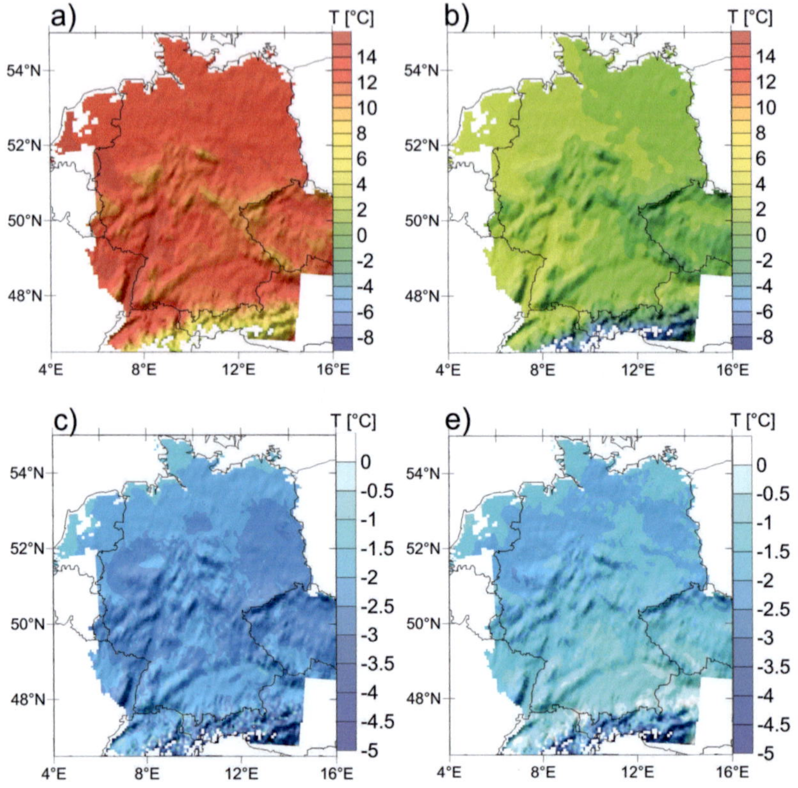

Figure 4.2: Climatological mean temperature [°C] in the reference period. Top row: Ensemble mean climatological means for a) summer b) winter, bottom row: mean difference between ensemble mean and HYRAS Observations for c) summer and d) winter.

an offset (bias) ranging between 0.8°C in the winter months and 2.8 °C in August. The ensemble spread lies between 0.8 and 1.5 °C for the mean monthly values. Depending on the region and for smaller time scales this might be larger. As to be expected, the temperature depends strongly on elevation.

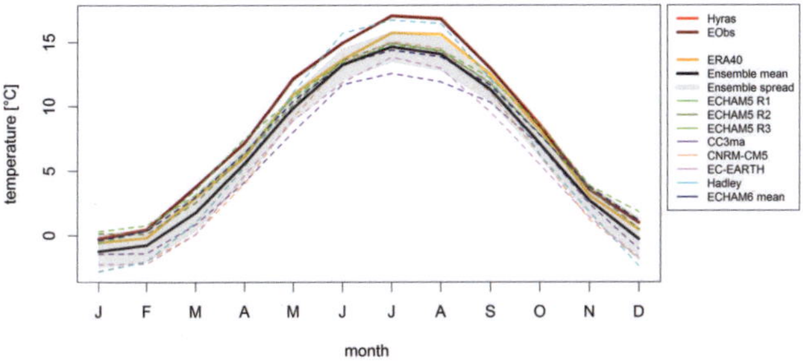

Figure 4.3: Climatological mean monthly temperature [°C] in the reference period. Areal mean over the HYRAS domain (see Fig. 4.1). Temperature means of HYRAS and E-Obs observations are indistinguishable.

Fig. 4.2 shows the ensemble mean of the climatological mean summer and winter temperatures. The mean summer temperatures range between -1 and 15°C, the winter mean temperatures between -11 and -5 °C. As already noted for the areal mean of the monthly mean temperature (Fig. 4.3), the bias is larger in summer than in winter. This cold bias of the COSMO-CLM in central Europe has already been observed in previous studies (e.g. Berg et al., 2013, 2012; Sedlmeier and Schädler, 2014). Berg et al. (2013) interpreted the temperature bias as an interaction between precipitation, cloudiness and temperature.

4.3 Maximum temperature

Since not all model runs had the maximum temperature as output variable, it is calculated as the maximum of the mean hourly temperature values. The 30 year mean of the monthly temperature maxima is shown in Fig. 4.5. As for the temperature means, the values follow the yearly cycle but show a negative bias compared to the E-Obs observations which is again larger in the summer months than in the winter. Mean winter and summer maxima are shown in Fig. 4.4.

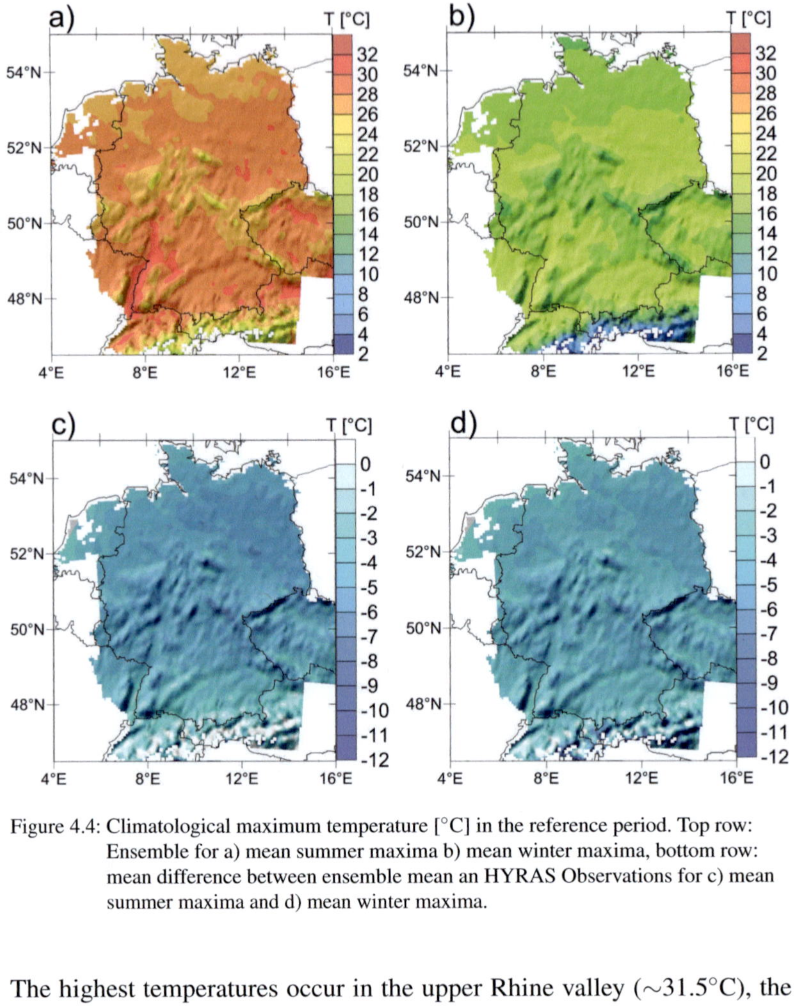

Figure 4.4: Climatological maximum temperature [°C] in the reference period. Top row: Ensemble for a) mean summer maxima b) mean winter maxima, bottom row: mean difference between ensemble mean an HYRAS Observations for c) mean summer maxima and d) mean winter maxima.

The highest temperatures occur in the upper Rhine valley (~31.5°C), the lowest in the Alps (~12.5°C). The areal mean over the HYRAS domain is ~27.5°C. Winter minimum temperatures lie between 0.5 and -21°C. Compared to the E-Obs mean temperature maxima (Fig. 4.4 c and d), these values are too low for both seasons - as Fig. 4.5 shows, there is a negative bias for all months.

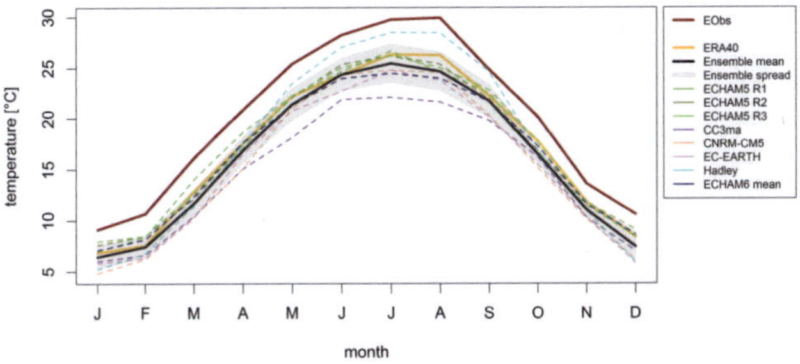

Figure 4.5: Climatological monthly maximum temperature [°C] in the reference period. Areal mean over the HYRAS domain (see Fig. 4.1).

4.4 Minimum temperature

Like the maximum temperature, the daily minimum temperature is calculated from the mean hourly temperature values. Fig. 4.6 shows the climatological monthly temperature minima.

The ensemble mean matches the E-Obs observations very well, the E-Obs yearly cycle lies within the ensemble spread for all months and correctly represents the maximum in July and the minimum in January for the average over the HYRAS domain. However, when looking at the bias on a grid point basis (see Fig. 4.7 c and d), some areas show a negative, some a positive bias

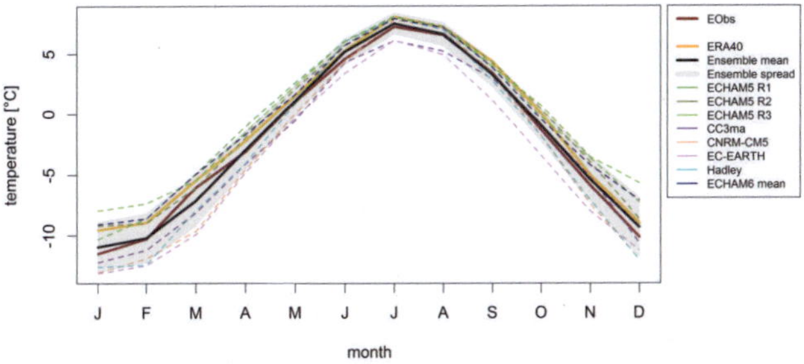

Figure 4.6: Climatological monthly minimum temperature [°C] in the reference period. Areal mean over the HYRAS domain (see Fig. 4.1).

which cancels out in the mean but most parts of the domain have no or only a very small bias. The mean temperature minima for summer lie between -16.5 and -3°C and between -27.5 and -7.5 in winter (see Fig. 4.7 a and b).

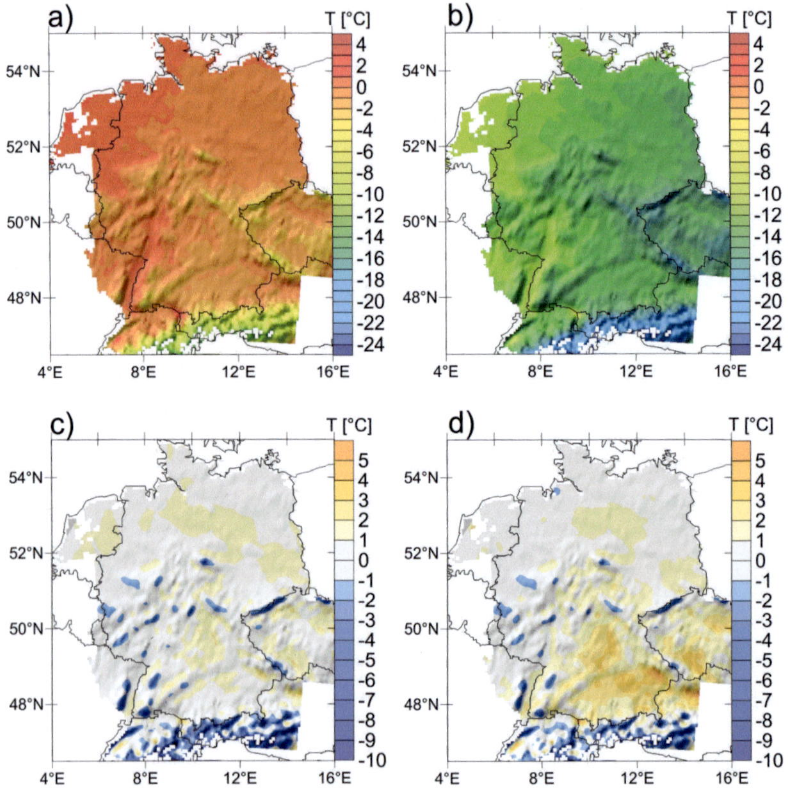

Figure 4.7: Climatological minimum temperature [°C] in the reference period. Top row: Ensemble mean for a) mean summer minima b) mean winter minima, bottom row: mean difference between ensemble mean and HYRAS Observations for c) mean summer minima d) mean winter minima.

4.5 Precipitation

Fig. 4.8 shows the climatological monthly precipitation sums. The ensemble mean can mostly represent the higher precipitation in summer and follows the annual cycle pretty well but shows an offset of up to 50 mm compared to the observations. The error is slightly higher in the winter months than in summer. The 30 year mean of the summer and winter sums of total precipitation is shown in Fig. 4.9. Regions with stronger orographic features show more precipitation, the maximum lies in the Alps. The precipitation mean for the whole HYRAS domain does not differ much between summer and winter and lies around 650 mm. Most parts of the domain show a positive bias which is especially high in the Alps and in winter (see Fig. 4.9 c and d). Previous studies have also shown a positive precipitation bias over central Europe (e.g. Berg et al., 2013, 2012; Sedlmeier and Schädler, 2014). Berg et al. (2013) attributed this bias to a correction of the mass loss in the cloud ice scheme, "which explains the stronger bias in winter than in summer". The bias is accompanied by a bias in shortwave radiation (Schädler et al., 2012; Sedlmeier and Schädler, 2014) due to an overestimation of the cloud cover in the COSMO-CLM (Will and Wold, 2009). There has been some effort of the COSMO-CLM community to reduce this bias.

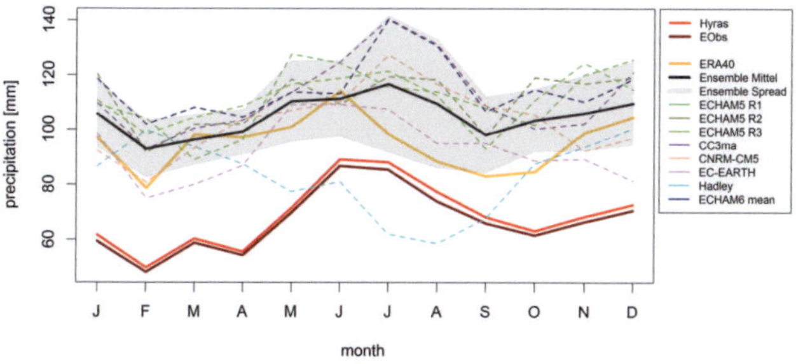

Figure 4.8: Climatological monthly precipitation sums [mm] in the reference period. Areal mean over the HYRAS domain (see Fig. 4.1).

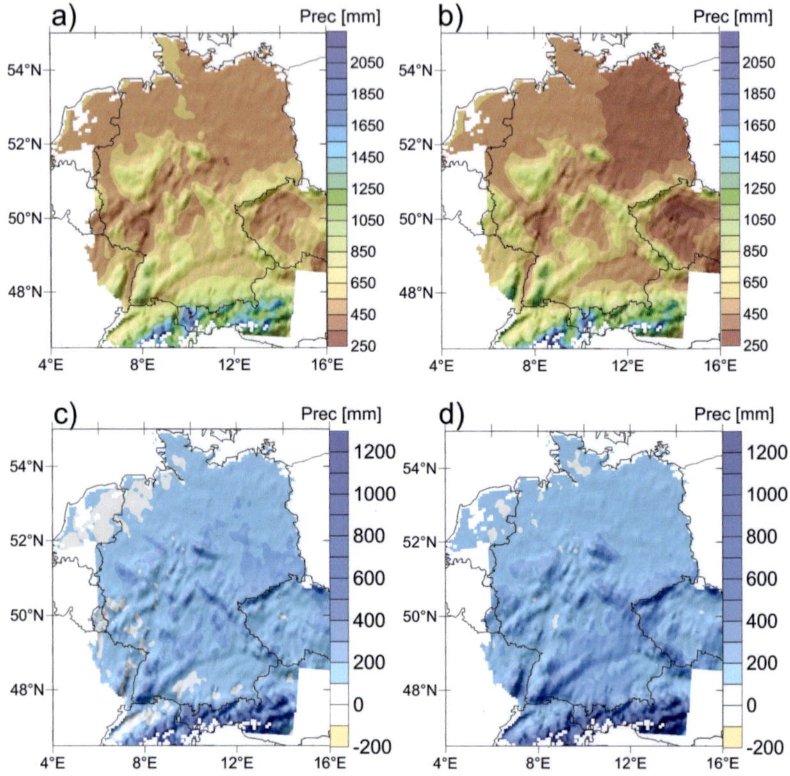

Figure 4.9: Precipitation sums [mm] in the reference period. Top row: ensemble mean climatological means for a) summer sums b) winter sums, bottom row: mean difference between ensemble mean an HYRAS Observations for c) summer sums and d) winter sums.

A project group formed to investigate this problem, however, did not gain any understanding as to the direct source of this error. The HYRAS observational dataset is also not free from biases. The average error of the HYRAS dataset is stated as being less than 2 mm a day but with spatial and temporal variability (Rauthe et al., 2013). Assuming a systematic error in one direction, monthly error bars could be as high as ~60 mm and half yearly error bars up to

360 mm. Considering these error bars, the ensemble performs fairly well for most of the region except for the Alps.

4.6 Added value of high resolution data

Downscaling of climate data comes with a higher computational demand and therefore usually with a reduction of the domain size. The driving idea behind downscaling is to obtain more detailed and better information about the regional climate. For a comparison of the two resolutions, the data of the 50 km ensemble was bilinearly interpolated to the 7 km grid and a height correction applied for temperature. Fig. 4.11 shows a comparison of the mean monthly temperature means for the reference period. The deviations from the HYRAS observations are slightly lower in winter and slightly higher in summer for the 7 km ensemble, but overall the biases do not differ much and the ensemble spreads nearly overlap. As a further measure of comparison, the spatial RMSE (root mean square error) of the mean summer/winter temperature means was calculated using all gridpoints within the HYRAS domain for four different elevation ranges (as shown in Fig. 4.10) and the correlation of gridpoints of all heights was assessed. Both measures were

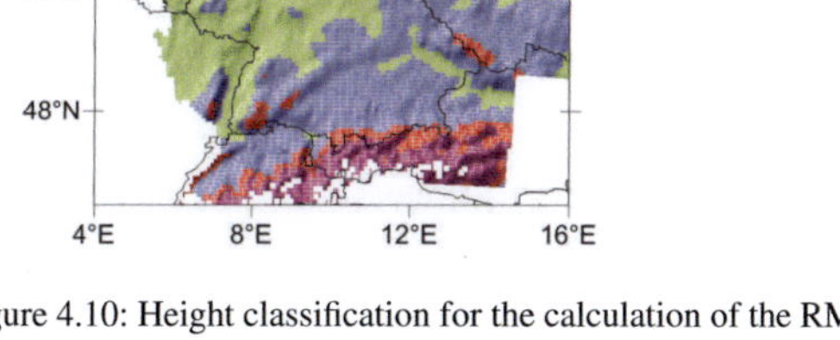

Figure 4.10: Height classification for the calculation of the RMSE

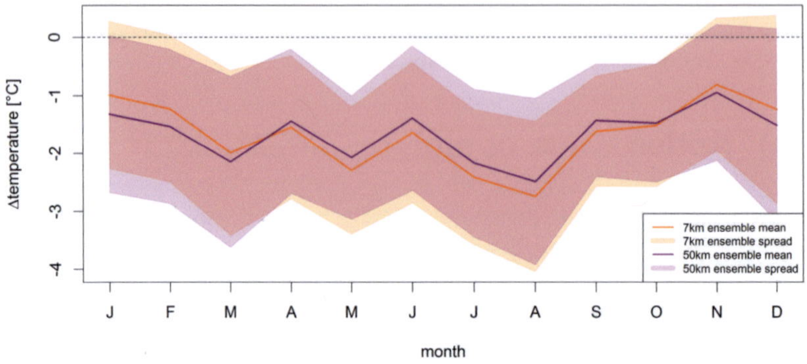

Figure 4.11: Difference of mean monthly temperature means between 7 km (orange) and 50 km (magenta) ensemble and HYRAS observations for the reference period (1971-2000). Lines: ensemble means, shaded area: ensemble spread. Areal mean over the HYRAS domain (see Fig. 4.1)

calculated for each ensemble member separately and in Fig. 4.12, 7 km and 50 km ensembles are compared by boxplots. They show the median and interquartile range and the whiskers indicate the minimum and maximum value of the respective ensemble. These results show tendencies similar to the monthly temperature means. For elevations below 800 m the errors are very similar. For heights between 800 m and 1200 m the 7 km ensemble is slightly better, and above 1200 m vice versa.

The mean values for the errors in summer are higher than in winter and the ensemble spread is smaller. For winter temperature biases the ensemble members exhibit a larger spread. The correlation is significantly higher for the 7 km ensemble for both seasons.

The same analysis was conducted for precipitation means. Here, the 50 km ensemble performs better regarding the biases. The ensemble mean bias for mean monthly precipitation sums (see Fig. 4.13) is around 10 mm higher for the 7 km ensemble in all months, although the ensemble spreads overlap. The RMSE for mean summer and winter sums for 4 different elevation regimes of the HYRAS domain is shown in Fig. 4.14. For elevations up to 800 m, the

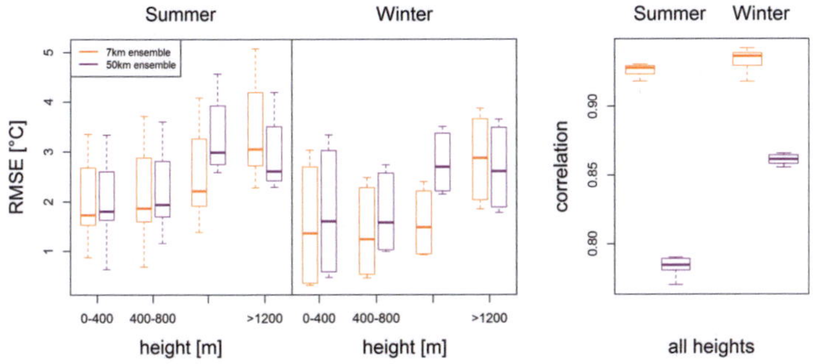

Figure 4.12: Root mean square error (RMSE) and spatial correlation for mean summer/winter temperatures means of the 7 km and 50 km ensembles with respect to HYRAS observations for the HYRAS domain. Gridpoints were grouped according to elevation for calculation of the RMSE (see Fig. 4.10)

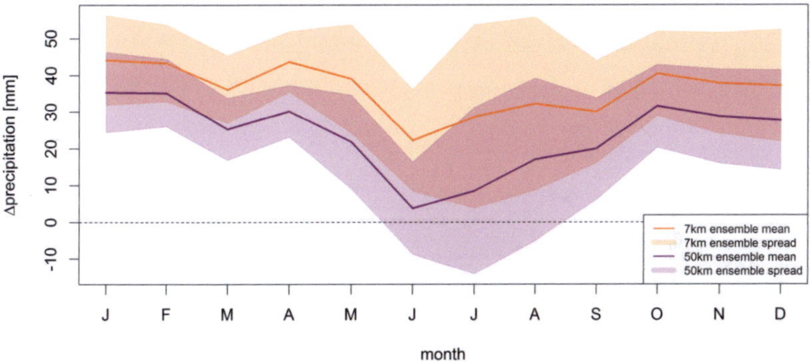

Figure 4.13: Difference of mean monthly precipitation sums between 7 km (orange) and 50 km (magenta) ensemble and HYRAS observations for the reference period (1971-2000). Lines: ensemble means, shaded area: ensemble spread. Areal mean over the HYRAS domain (see Fig. 4.1)

RMSE is fairly similar for ensembles of both resolutions. Above 800 m the RMSE of the 7 km ensemble is much higher and the second nesting stage seems to significantly increase the ensemble spread. The correlation is, however, higher for the 7 km ensemble for both seasons, although the difference

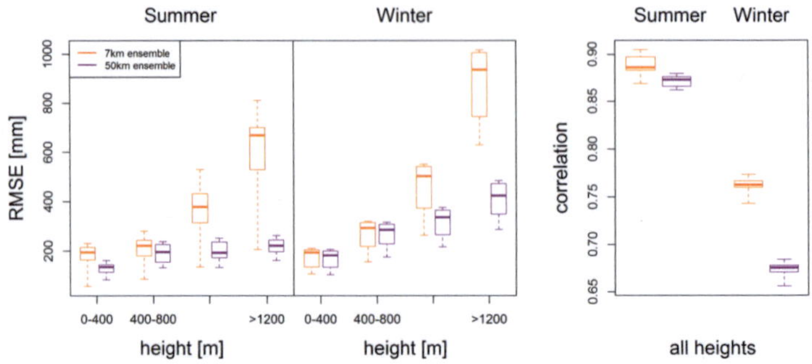

Figure 4.14: Root mean square error (RMSE) and spatial correlation for mean summer/winter precipitation sums of the 7 km and 50 km ensembles with respect to HYRAS observations for the HYRAS domain. Gridpoints were grouped according to elevation for calculation of the RMSE (see Fig. 4.10)

is only significant in winter. A possible reason for the overestimation of precipitation might be the parametrization of convection. For 50 km resolution simulations, parametrization of convection is clearly needed. Starting at resolutions of below about 3km, the convective processes can be explicitly resolved by the model. At 7 km resolution, the parametrization of convection is still turned on in the model, but it might already be able to resolve some processes which are then accounted for twice. The parametrization of convection are most probably tuned to coarser resolutions. A further error source is the observational dataset, which shows the highest uncertainties in these high elevation regions.

4.7 Short summary

This chapter gave an overview of the performance of the 7 km ensemble for the values which are used later for the calculation of extreme values, namely mean, maximum and minimum temperature and precipitation sums for the summer and winter season. The 7 km ensemble shows a cold bias in mean and maximum temperature (with a larger bias in maximum than in mean values) for the HYRAS domain. The minimum temperature has a warm bias, mainly in southern Germany in Winter, the amplitude of the daily cycle seems to be underestimated here. The Alps and some other high elevation regions show a cold bias. The ensemble mean precipitation sum shows a wet bias throughout the year and for the whole domain. Comparing these results to the first nesting stage, the 50 km ensemble, there is an added value in the representation of spatial patterns of temperature and precipitation. However, the 7 km ensemble seems to overestimate the precipitation more strongly than the 50 km ensemble, especially in regions with higher elevation.

5 Bias correction of ensemble members

No climate model is perfect and model data generally exhibits a bias in comparison to observational data. This can have different causes, a bias in the GCM data used as boundary condition, insufficiently high resolution, deficiencies in parametrizations, missing comprehension of the underlying processes or uncertainties in the observational datasets. Whenever working with absolute values or when using the output of regional models as driving data for impact models, this bias becomes problematic and a bias correction is unavoidable. Generally, there are two different approaches to correct data from RCMs. The first one is the correction of the GCM data used as initial and boundary condition (e.g. , Colette et al., 2012; Bruyère et al., 2014). This of course still leads to the bias of the RCM itself but could make a change from "garbage in - garbage out" to "not garbage in — not garbage out". The second approach is to bias correct the RCM output itself, for which numerous methods exist (e.g. , Berg et al., 2012; Hoffmann and Rath, 2012; Teutschbein and Seibert, 2012). The great advantage of the first method is the physical consistency of the results. When bias correcting model output usually only single variables or a set of variables are corrected. It is not possible to correct all variables due to the lack or scarcity of observational data for the required time periods. This could for example lead to days where the model shows no cloud cover but after correcting the precipitation data it is raining. The advantage of the correction of the regional output is that it is by far the simpler and less time-consuming alternative, especially when statistical evaluation of only some variables is planned. Furthermore, the bias of the

RCM is still present when correcting the global climate data. Depending on the magnitude of the bias, the regional output would still have to be corrected for some applications. Therefore in this work only the regional output data was bias corrected.
For mean values of temperature and precipitation, the linear single scaling method (Section 5.1) was used, for maximum and minimum temperature a quantile mapping approach (Section 5.2). A detailed explanation and discussion can be found in Berg et al. (2012), where different bias correction methods were applied to two members of the ensemble used in this work. An underlying assumption for all bias correction techniques is the persistence of the bias, meaning that the bias in future time periods has the same statistical values as in the reference period and can be corrected using the same correction term. This of course is an assumption which is certainly not always true and depends largely on the nature of the bias. Furthermore the bias correction affects the climate change signal of the corrected variables. Examples of this are given in Section 5.3.

5.1 Linear bias correction

With the linear single scaling (LSS) method, the model data is bias corrected by addition or multiplication of a correction term such that the monthly climatological values match those of the observations for the reference period, 1971-2000. The correction terms are deduced from the mean monthly temperature means and mean monthly precipitation sums, respectively. The values for the reference time period (1971-2000) are applied to the future time period as well. The correction is performed separately for each gridpoint and each model run and the HYRAS dataset used for the correction.

5.1.1 Linear correction of temperature

The first step consists of calculating the climatological monthly temperature means for the reference time period (1971-2000) for the observations $\overline{T^{O}_{m,g}}$ and model data $\overline{T^{M}_{m,g}}$ (m=months 1-12, g=gridpoint). The difference in climatological mean values ($\Delta T_{m,g} = \overline{T^{M}_{m,g}} - \overline{T^{O}_{m,g}}$) constitutes the climatological monthly correction term for each gridpoint. The bias corrected daily temperature values are the difference between the model data on a day d of month m and the correction term for the corresponding month:

$$T^{M,BC}_{m,g,d} = T^{M}_{m,g,d} - \Delta T_{m,g}$$

5.1.2 Linear correction of precipitation

The COSMO-CLM usually shows too few dry days (days with precipitation less than 1 mm/day) in the investigation area (Feldmann et al., 2008). Therefore, an additional step is inserted before the linear correction to correct the number of dry days.

1. correction of dry days: To correct this deviation from the observations, the mean number of dry days per month is evaluated for each gridpoint for the reference period (1971-2000). In a next step, the daily precipitation sums of the model data are sorted in ascending order and the precipitation threshold value identified which leads to the same number of dry days as for the observations. Then, the actual values are reduced by this threshold value, possible negative values are set to zero.

2. In a second step, the climatological monthly precipitation sums are calculated for the reference period (1971-2000) for the observations $\overline{PREC^{B}_{m,g}}$ and the dry day corrected model data $\overline{PREC^{M}_{m,g}}$ (m = months 1-12 ; g = gridpoint). The quotient α of the mean precipitation sums constitutes the climatological monthly correction factor for each grid-

point. The bias corrected precipitation values are obtained by multiplying the dry day corrected model data of every day d with the correction factor α of the corresponding month:

$$PREC_{m,g,d}^{M,BC} = \alpha * PREC_{m,g,d}^{M}; \alpha = \frac{\overline{PREC_{m,g}^{B}}}{\overline{PREC_{m,g}^{M}}}$$

5.2 Quantile mapping

The quantile mapping, or histogram equalization method is applied to minimum and maximum temperature. In addition to the mean, it also corrects the second order moment (the variance) of the distribution. In the following, the method is shortly summarized, a more detailed description can be found in Berg et al. (2012). For this bias correction method the lengths of the time series to be corrected and the observational time series need to be the same. In a first step, both time series are sorted in ascending order and plotted against each other (in a Q-Q plot). If the model data were to have no bias at all, that is, if both datasets were exactly the same, they would lie on the x=y line. Deviations from this line indicate a bias in the results. If the resulting curve lies under the x=y line, the model results show a positive bias, if they are over the line they exhibit a negative bias. The bias correction method is based on deriving an empirical transfer function which projects the curve of observations versus model results onto the x=y line. There are several methods of estimating this transfer function, in this work a linear fit was used. By this, moments higher than the second order are not corrected. The method was applied to monthly data for the 30 years of the reference period, i.e. data from the Januaries of each of the 30 years and, thus, different transfer functions were estimated for each months and gridpoint. Transfer functions from the reference period were then used to bias correct the data of the future time period. Since the HYRAS dataset does not include minimum and maximum daily temperature, E-Obs data was used which was bilinearly interpolated to the model grid and height corrected (see Section 4.1.2).

5.3 Effect of bias correction on the climate change signal

In most cases, bias correction affects the climate change signal of the corrected variables. Depending on the method and the temporal resolution, the difference between uncorrected and corrected climate change signal varies in magnitude. In Fig. 5.1 the climate change signal for variables corrected as described in the preceding sections is exemplarily compared to the uncorrected change signal for one ensemble member (COSMO-CLM driven by the GCM ECHAM6). For temperature means corrected according to Section 5.1.1, the climate change signal of mean values does not change, since the additive correction factor cancels out (top left figure). This also holds for different temporal resolutions. For a multiplicative correction, the relative change signal would be the same (as the multiplicative factor would cancel out). For the linearly corrected precipitation, however (correction described in Section 5.1.2), the number of dry days was additionally corrected leading to a change in relative and absolute change signal. When looking at the same temporal resolution as the correction factors (mean monthly sums), the direction of change is the same, but the magnitude of the change signal differs between the months, depending on the bias to observations.

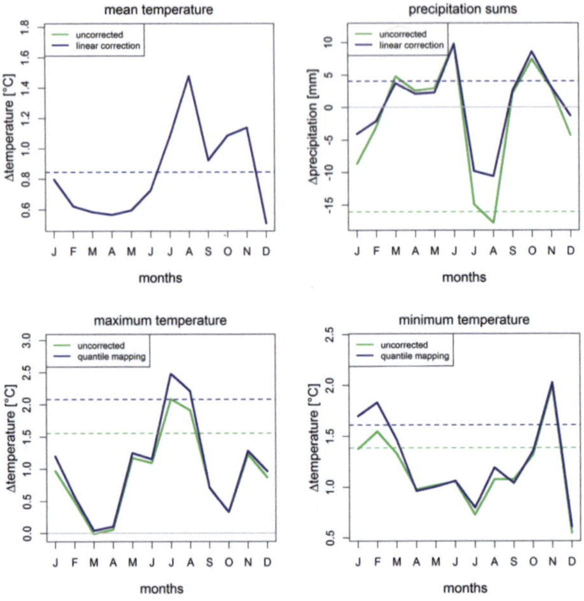

Figure 5.1: Climate change signal between reference period (1971-2000) and near future (2021-2050) for bias corrected versus non bias corrected temperature and precipitation variables for one selected ensemble member (see text). Solid lines: mean monthly temperature means/minima/maxima and mean monthly precipitation sums, dotted lines: mean yearly temperature means/minima/maxima and mean yearly precipitation sums. Area mean over HYRAS domain (see Fig. 4.1)

When calculating the mean yearly sums, however, corrected and uncorrected climate change signals even have different signs - the corrected precipitation shows a slight decrease whereas the uncorrected shows an increase. However, the magnitude of the difference is rather small compared to the total precipitation amount and generally no significant changes are expected for precipitation (see Chapter 6).

For the temperature minima and maxima corrected with the quantile mapping method (see Section 5.2), the climate change signal is also altered to a higher value. The maximum temperature shows a larger cold bias with respect to

observations (see Chapter 4) and, therefore, the effect on the change signal is also greater. One of the reasons for bias correcting data is related to the calculation of extremes defined as threshold exceedances.The bias correction does not only affect changes in mean values but also in statistical parameters calculated from the bias corrected daily time series. As an example the number of hot days (days with a maximum temperature above 30°C) for the ECHAM6 model and E-Obs observations are shown in Fig. 5.2.
The bias corrected data overestimate the number of hot days per year, but are closer to the observed values than the uncorrected data. One reason for this overestimation could be an over-correction because the maximum daily temperature of the ensemble was calculated as the maximum of hourly means (see Chapter 4.3) and the actual maximum temperature values could be higher. Another reason for an over-correction is a skewed distribution, where a linear estimation of the transfer function can lead to an overestimation at the tails of the distribution (Berg et al., 2012). The climate change signal of hot days between the two time periods is more than five times as higher for bias corrected maximum temperature series than for the uncorrected data.

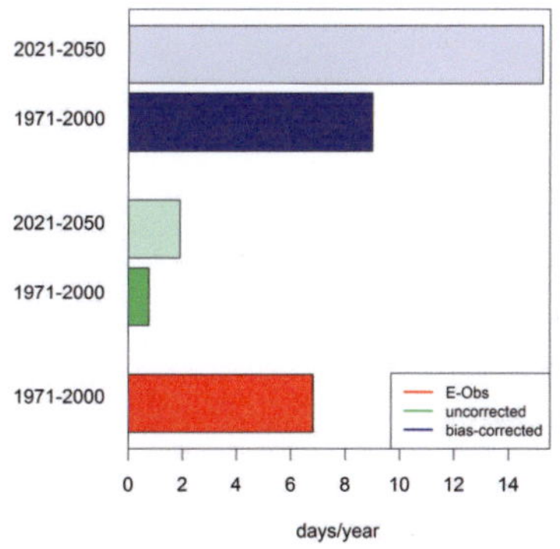

Figure 5.2: Number of hot days per year for a gridpoint in the Rhine Valley for one selcted ensemble member (see text). Values for reference and future time periods for uncorrected and bias corrected data as well as for E-Obs Observations.

Bias correction of RCM output has an effect on the climate change signal. The magnitude of this effect depends on the bias correction method, the statistics considered (means, sums or threshold exceedance) as well as the temporal resolution of the analysis. This needs to be kept in mind when working with bias corrected data.

5.4 Short summary

Due to the cold and wet bias of the 7 km ensemble mean, temperature and precipitation data were bias corrected using a linear method (for daily temperature means and precipitation sums) and quantile mapping (for minimum and maximum temperature). These bias corrected values are used in Chapters 6 and 7. Whenever using bias corrected data, it needs to be kept in mind that the bias correction also affects the climate change signal in most cases.

6 Changes of ensemble mean values

In the following sections, the change signal for mean, minimum and maximum temperature and precipitation sums between the reference time 1971-2000 and the near future 2021-2050 is derived for the 7 km ensemble. All results are based on bias corrected data (see Chapter 5). Shown are the ensemble mean and spread of climatological monthly means/sums and summer/winter means/sums. As a measure of robustness of the data, the ensemble consistency (see Section 2.3) and the significance (see Section A) of the change signal were calculated.

6.1 Mean temperature

The changes in climatological mean temperatures are depicted in Fig. 6.1. Except for one model in the month of April, all models show an increase for

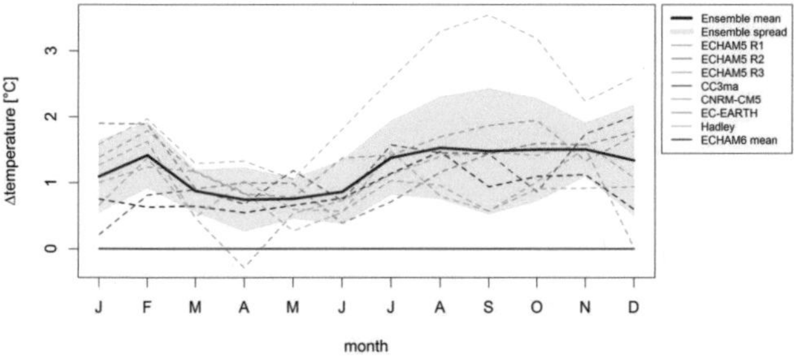

Figure 6.1: Changes in climatological mean monthly temperature [°C] between 1971-2000 and 2021-2050. Areal mean over the HYRAS domain (see Fig. 4.1).

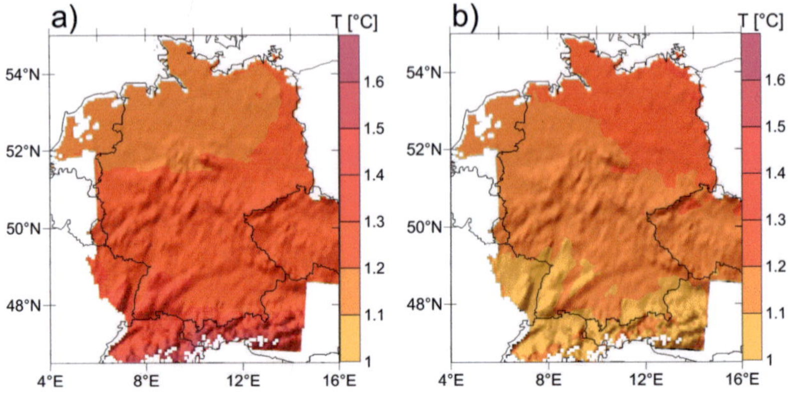

Figure 6.2: Changes in climatological mean temperature [°C] between 1971-2000 and 2021-2050 for a) summer b) winter. The ensemble consistency is 100% for the whole investigation area and therefore not shown.

all months, the highest increases being in August through November. Since the ensemble spread is always above the "zero line", the ensemble shows a significant change in temperature. This is in accordance with findings of other studies (Jacob et al., 2014; IPCC, 2013).
The mean changes in summer (between 1.1 and 1.7 °C) are slightly higher than in winter (between 1.0 and 1.3 °C). In summer, the change is higher in the Alps and Southern and Central Germany whereas winter changes are highest in the northeastern part of the HYRAS-domain (see Fig. 6.2). The changes are significant at the 5% level for all gridpoints and the ensemble consistency of the change signal is 100 % for all gridpoints and summer as well as winter and is therefore not shown.

6.2 Maximum temperature

The changes in the mean monthly temperature maxima are depicted in Fig. 6.4. As for mean temperatures, except for one model in April, all models show an increase for all months. The temperature maxima show a stronger increase than the means with a maximum of 2.2°C in August. Changes in summer are stronger than in winter (Fig. 6.3). In winter, the Alpine region shows particularly high changes. The changes are significant for all but a few gridpoints in winter (mostly in the western part of the model domain in France) and the ensemble consistency of the change signal (Fig. 6.3 c and d) is 50% or higher for all regions.

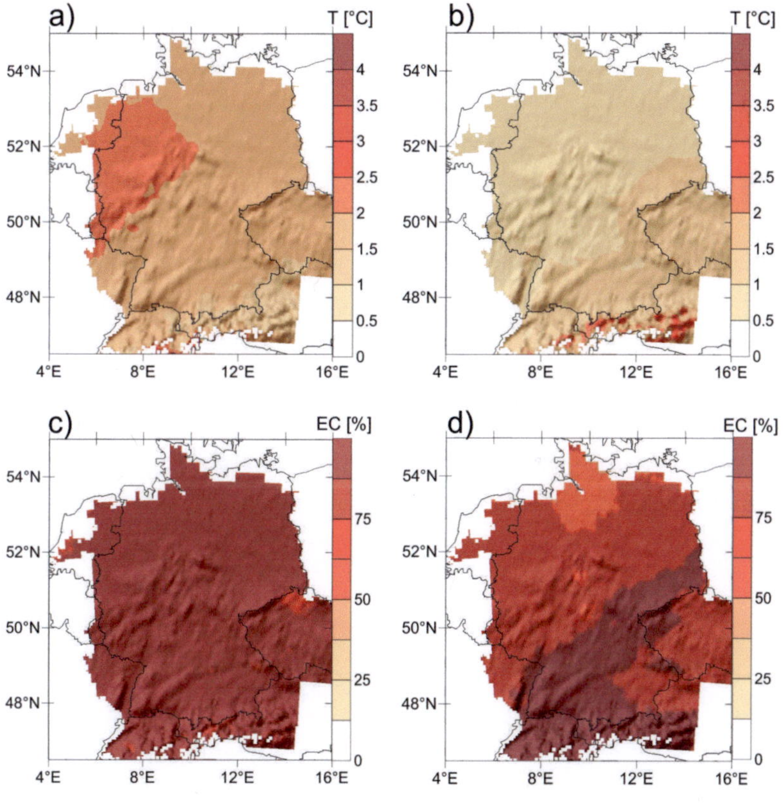

Figure 6.3: Changes in climatological maximum temperature [°C] between 1971-2000 and 2021-2050. Top row: Changes of ensemble mean climatological maxima for a) summer b) winter, bottom row: ensemble consistency of change signal for c) summer d) winter. The change signal is significant at the 5 % level for more than 99 % of the area.

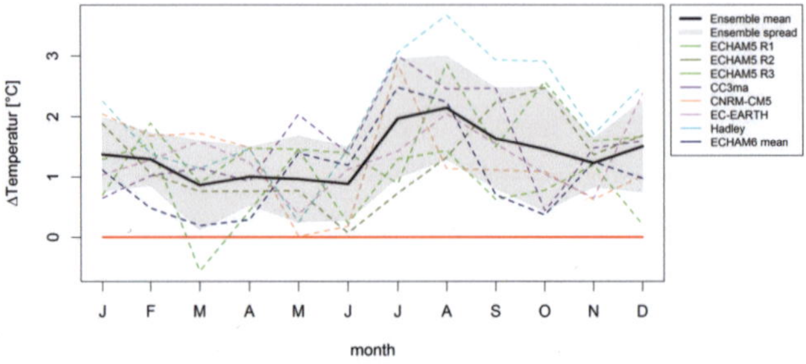

Figure 6.4: Changes in climatological monthly maximum temperature [°C] between 1971-2000 and 2021-2050. Areal mean over the HYRAS domain (see Fig. 4.1).

6.3 Minimum temperature

The change signal of the mean monthly temperature minima is depicted in Fig. 6.5. Again, the ensemble spread of the change signal clearly lies above the "zero line" and the ensemble projects a significant increase for all months. The peaks of increase are in February (2.4°C) and November (2.1°C), generally the increase is higher in winter than in summer. The temperature increase in summer lies between 0.9 and 2.3°C and between 0.8 and 3.1°C in winter.

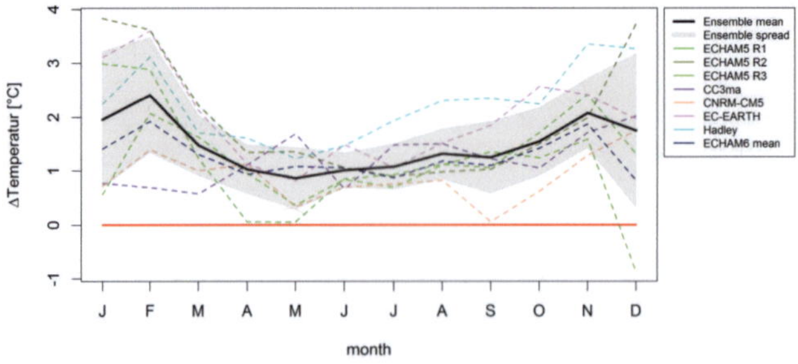

Figure 6.5: Changes in climatological monthly minimum temperature [°C] between 1971-2000 and 2021-2050. Areal mean over the HYRAS domain (see Fig. 4.1).

The ensemble consistency (Fig. 6.6 c and d) is above 50% for the whole model domain and the changes are significant for all but a few gridpoints in the Alps in winter. Minimum temperatures change slightly stronger than maximum temperatures, this has also been observed for past records of observational data (e.g. Heino et al., 1999; Beniston et al., 1994).

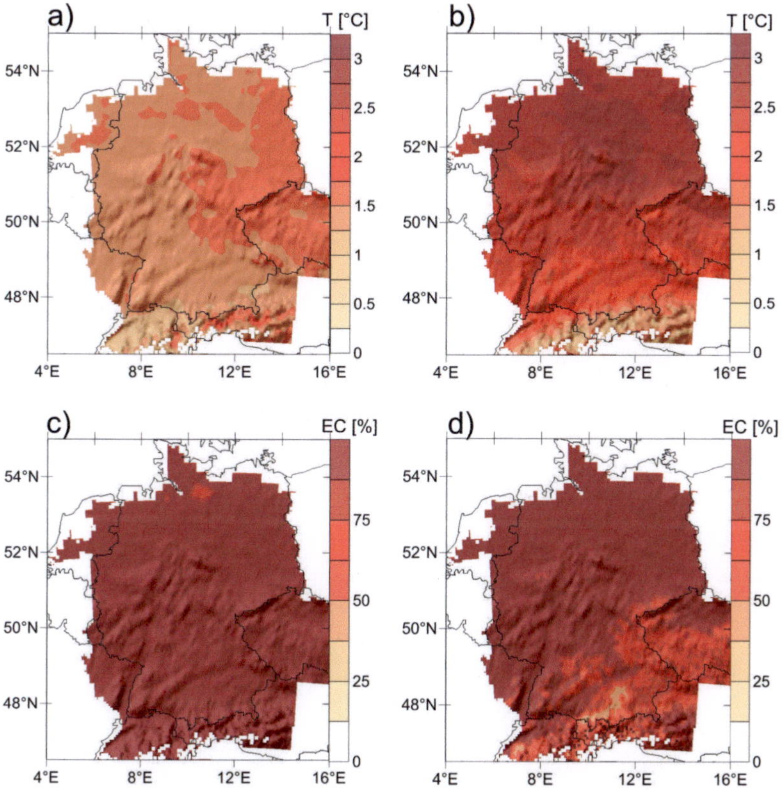

Figure 6.6: Changes in climatological minimum temperature [°C] between 1971-2000 and 2021-2050. Top row: Changes of ensemble mean climatological minima for a) summer b) winter, bottom row: ensemble consistency of change signal for c) summer d) winter, non significant changes are marked with a cross. The change signal is significant at the 5 % level for more than 99 % of the area.

6.4 Precipitation

Precipitation shows a clear change signal only in some regions. The mean of the climatological monthly sums for the HYRAS-domain (Fig. 6.8) shows an increase of precipitation in most months with a negative peak and, thus, a decrease in the months of July and August (by 8.8/8.3 mm). The "zero line"

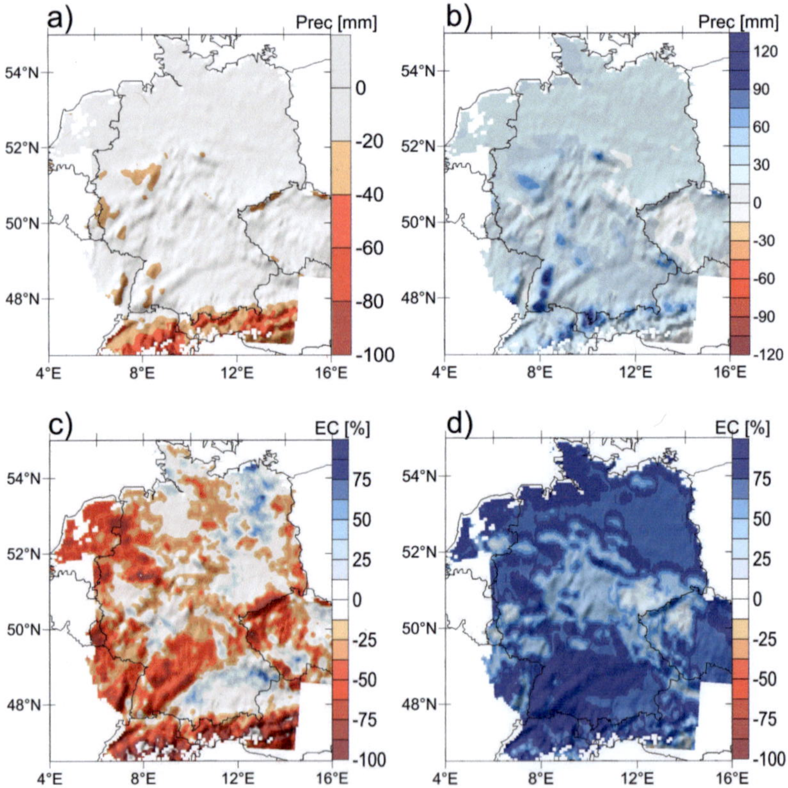

Figure 6.7: Changes in climatological monthly precipitation sums [mm] between 1971-2000 and 2021-2050. Top row: Changes of ensemble mean climatological sums for a) summer b) winter, bottom row: ensemble consistency of change signal for c) summer d) winter. The change signal is not significant for all gridpoints in summer and winter.

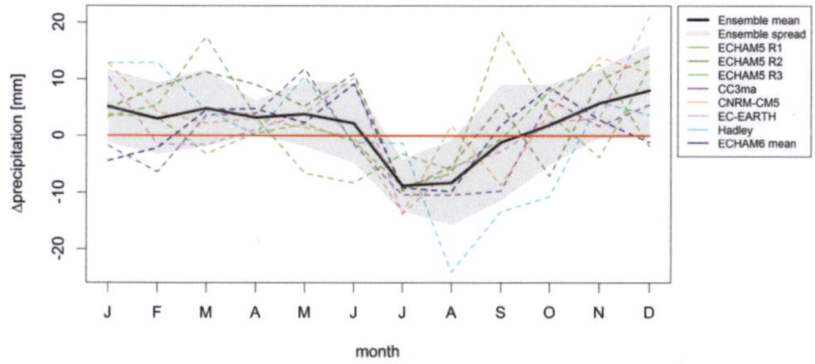

Figure 6.8: Changes in climatological precipitation sums [mm] between 1971-2000 and 2021-2050. Areal mean over the HYRAS domain (see Fig. 4.1).

goes through the ensemble spread for all months, there is no clear change signal.

Figs. 6.7 a and b show the areal mean winter and summer precipitation sums. In summer, the ensemble predicts a decrease for the Black Forest and western Germany as well as the Alps (up to 90 mm), the rest of the investigation area shows only small changes with a low ensemble consistency (see Fig. 6.7 c). The change signal is not significant for the whole investigation area. In winter, there is a slight increase for most of the investigation area with especially high values in the Black Forest and the Alps. For winter increase, the ensemble consistency is also fairly high for most part of the region. These findings also agree with the general tendencies found in other studies (e.g., Jacob et al., 2014; IPCC, 2013).

6.5 Short summary

This chapter gave an overview of the change signal of mean, maximum and minimum temperature and precipitation sums for the summer and winter season between the recent past (1971-2000) and the near future (2021-2050). The 7 km ensemble used in this work shows a robust increase in mean, maximum and minimum temperature values for central Europe. Changes in minimum temperatures are higher than for maximum temperatures. This increase is significant except for a few gridpoints in the Alps for minimum and maximum temperature. For precipitation, most ensemble members show a small increase in winter and a decrease in summer. However, these changes are not significant.

7 Compound extreme events with thresholds: type 1 extremes

This chapter contains an analysis of compound extreme events using thresholds, namely compound hot and dry extremes in summer (JJA) and compound cold and heavy precipitation days in winter (DJF). A compound event is defined by both temperature and precipitation exceeding a certain threshold on a given day. The following thresholds are used:

Table 7.1: Thresholds for daily temperature and precipitation events.

hot extreme	maximum daily temperature > 30°C
cold day	minimum daily temperature < 0°C
dry day	daily precipitation sum < 1 mm
wet extreme	daily precipitation sum > 25 mm

These extremes play a role for agriculture and infrastructure among others. The statistical parameters considered for the analysis are the number of compound extreme days, the number of compound extreme episodes and the mean length of episodes. For the latter two, an episode is defined as at least two consecutive compound extreme days. All measures are given on a mean yearly basis and calculated from bias corrected data (as described in Chapter 5).

7.1 Hot and dry extremes in summer

Table 7.2 gives an overview of the ensemble mean of the statistical parameters of univariate hot/dry days and compound hot and dry days in summer

(JJA) for the reference period (1971-2000) and the climate change signal (1971-2000 vs. 2021-2050). The minimum, maximum and areal mean of the HYRAS domain are listed for the number of extreme days and episodes and the mean episode length. Note that the minimum/maximum value of the reference period and the change signal are most likely not at the same grid point. The results for compound extreme events are depicted graphically in the following two sections. For validation, the E-Obs dataset was used (see Section 4.1.2), as the HYRAS dataset does not contain values for daily minimum and maximum temperature and for reasons of consistency both variables were taken from the same observational dataset.

7.1.1 Validation

The statistical parameters for compound hot and dry extremes in summer for the reference period are shown in Fig. 7.1. The number of compound hot and dry days exhibits a dependence on elevation; the Alps and the lower mountain ranges show no or only a small number of extreme days.

Table 7.2: Mean, minimum and maximum values of all gridpoints of the HYRAS domain of different statistics for the reference period (1971) and the climate change signal (1971-2000 vs. 2021-2050). Comparison between univariate hot/dry and compound hot and dry extremes in summer (JJA).

		dry		**hot**		**hot and dry**	
statistic		**ref**	**change**	**ref**	**change**	**ref**	**change**
number	**min**	37.9	1.4	0	0	0	0
of	**max**	69.1	5.4	14.0	8.8	11.9	7.5
days	**mean**	58.8	3.2	4.9	4.7	4.1	4.0
number	**min**	8.6	-1.3	0	0	0	0
of	**max**	12.3	0.6	3.1	1.7	2.7	1.7
episodes	**mean**	11.0	-0.4	1.1	1.0	0.9	0.9
mean	**min**	3.5	0.18	0	-0.8	0	-0.8
episode	**max**	6.3	1.13	3.8	1.7	3.5	1.8
length	**mean**	4.8	0.6	2.8	0.4	2.7	0.4

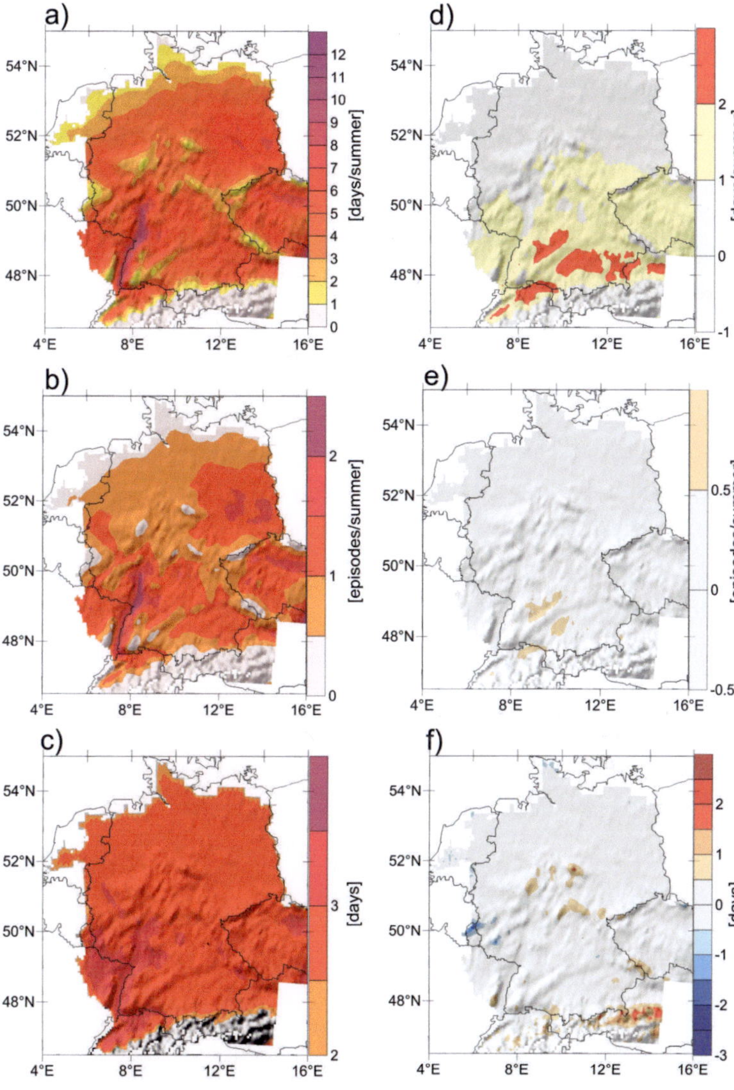

Figure 7.1: Compound hot and dry extremes in summer (JJA) in the reference period (1971-2000). Left side: ensemble mean statistics ; right side: difference ensemble mean – E-Obs observations a)+d) mean number of days/summer, b)+e) mean number of episodes/summer, c)+f) mean episode length.

The highest number of hot and dry type 1 extremes occurs in the Rhine Valley and parts of Brandenburg, the coastal areas are less affected. The number of episodes and the mean episode length (Figs. 7.1 b, c) roughly match the distribution of the number of days (Fig. 7.1 a), the highest values are in the Rhine Valley, whereas in the Alps and along the northern coast there are less than 0.5 episodes per year but with a similar length as in the rest of the area. When comparing the mean, minimum and mean values of the HYRAS domain for all three statistical parameters (Table 7.2), it becomes clear that the occurrence of compound extreme hot and dry days is governed by the occurrence of hot days. The mean number of dry days is approximately one order of magnitude higher than that of hot days so these are mostly the limiting factor. The spatial pattern of Figs. 7.1 a-c also matches that of univariate hot days (not shown). A comparison to E-Obs observations (right side of Fig. 7.1) shows that the ensemble mean does very well in the northern part of the investigation area. In the southern part there are slightly too many hot and dry extremes and the mean length of the episodes is somewhat too high, but the number of episodes and the mean episodes length fit very well for almost all of the domain. The bias which is present for the number of days does not transfer to the other two statistics. The reason for this is that only episodes of at least 2 days lengths are considered and the deviation is largest for one day events. Possible reasons for the positive bias in the number of hot days, which leads to the positive bias in the number of compound hot and dry days, are discussed in Section 5.3.

7.1.2 Climate change signal

The change signal of the number of compound hot and dry episodes between the reference period and the near future is shown in Figs. 7.2 a - c. A comparison to the statistical parameters of the reference period (Figs. 7.1 a - c) shows that regions that show a high number of extreme events, like the Rhine Valley or regions in Brandenburg, are also prone to strong changes in the near future (up to 7.5 days per year). The coastal areas, which show a low number of extremes in the reference period, also yield a smaller absolute change signal in the near future (2-3 days per year). However, the relative changes in these regions are above 100%. Like the values for the reference period, the change signals of the number of days and the number of episodes are again correlated, whereas the mean length of episodes is again fairly homogeneous (between 0.25 and 0.5 days) for most of the domain.

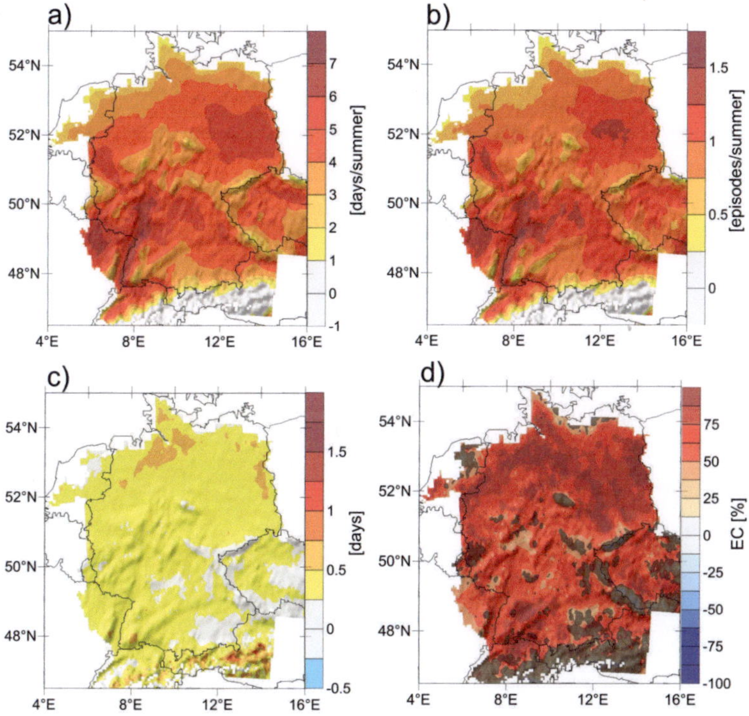

Figure 7.2: Ensemble mean climate change signal of compound hot and dry extremes in summer (JJA) between 1971-2000 and 2021-2050. a) mean number of days/summer, b) mean number of episodes/summer, c) mean episode length d) ensemble consistency of c, grid points where the change signal is not significant are shaded. EC of a and b is 100% for most of the investigation area and not shown (see text)

Just like in the reference period, the change signal is dominated by the change of hot days (see Table 7.2). Averaged over the whole HYRAS domain, the number of compound hot and dry days increases by more than 100%. As a measure of robustness, the ensemble consistency (see Section 2.3) and the significance of the change signal (see Section A) were calculated. These show a robust change in the number of compound extreme days and episodes and are therefore not shown; the ensemble consistency is 100% except for a few grid points in the Alps and the changes are significant for 95% of the HYRAS domain. The ensemble consistency of the mean episode length (Fig. 7.2 d) is above 50% for regions with a strong change signal and these changes are mostly also significant. The change signal of the ensemble for univariate extremes (hot days and dry days) is in rough agreement with those of other projects for this region (Beniston et al., 2007; Van der Linden and Mitchell, 2009; Jacob et al., 2014; Sedlmeier and Schädler, 2014) and the increase in hot days agrees with the increasing temperature trend (see Fig. 3.2). Thus, one can have confidence in these findings for compound extreme events.

7.2 Cold and wet extremes in winter

Table 7.3 gives an overview of the ensemble mean of the statistical parameters of univariate cold/wet days and compound cold and wet days in winter (DJF). The minimum, maximum and areal mean of the HYRAS domain are listed

Table 7.3: Mean minimum and maximum values of the number of univariate wet/cold and compound wet and cold extremes in winter (DJF) for the HYRAS domain, reference period (1971) and climate change signal (1971-2000 vs. 2021-2050).

		wet		cold		wet and cold	
statistic		**reference**	**change**	**reference**	**change**	**reference**	**change**
number	**min**	0	1.34	26.42	-14.37	0	-1.16
of	**max**	10.02	-0.12	90.37	-0.20	5.36	0.73
days	**mean**	0.41	0.11	58.3	-10.40	0.20	0.01

for the number of extreme days. Note that the minimum/maximum value of the reference period and the change signal are most likely not at the same grid point (i.e. the minimum change signal of -1.16 compound extreme cold and wet days is not for the same grid point which shows the minimum value (0) for the reference period). The results for compound extreme events are depicted graphically and explained in the following two sections. Since the number of compound extreme episodes and the mean episode length are so small and this type of extremes only occurs in higher elevated, mountainous terrain like the Alps, the Black Forest and the Vosges mountains, only the number of extreme days is shown.

7.2.1 Validation

The occurrence of the combination of cold days and heavy precipitation in winter for the time period 1971-2000 is shown in Fig. 7.3 a. These compound events are largely dominated by the occurrence of heavy precipitation days since their number is limited in winter, whereas the number of cold days is about one order of magnitude higher (see Table 7.3). In comparison to the number of compound extreme days calculated from the E-Obs dataset (Fig. 7.3 b), the value calculated from the 7 km ensemble shows a positive bias in large parts of the Alps and the Black Forest. This bias is due to a bias in the heavy precipitation events. However, in comparison to HYRAS precipitation data, the E-Obs dataset shows too few heavy precipitation events in mountainous terrain (Rauthe et al., 2013). Therefore, the bias between the ensemble mean and the E-Obs dataset is not only due to a bad representation of the ensemble values but also caused by a bias in the E-Obs data. Comparing heavy precipitation events to HYRAS data also yields a small positive bias but of much lower magnitude than when compared to E-Obs heavy precipitation events (not shown).

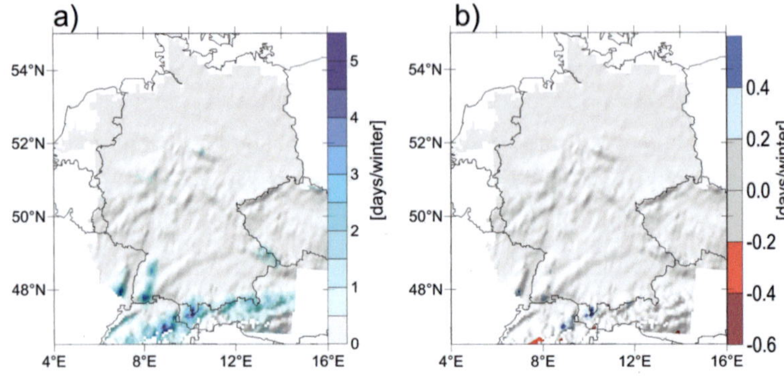

Figure 7.3: Ensemble mean compound cold and wet extreme days per year in winter (DJF) in the reference period (1971-2000).a) number of days/winter, b)difference ensemble mean – E-Obs observations.

7.2.2 Climate change signal

The 7 km ensemble does not show changes in most of the investigation area. Exceptions are mountainous regions (Fig. 7.4 a). The ensemble mean shows a decrease for parts of the Black Forest and the Vosges (up to $\approx 32\%$) and an increase in the Alps (by about 20%). The ensemble consistency (Fig. 7.4 b) is high in some areas, positive in the southeastern part of the HYRAS domain and negative in the southwestern part of domain. These areas also show significant changes but the change signal itself is negligibly small. The increase in the Vosges and the Black Forest shows a high ensemble consistency but the changes are not significant. These studies are in agreement with previous studies on univariate extreme events which show a decreasing trend in cold days (e.g. Heino et al., 1999) and an increase in heavy precipitation days in mountainous regions in winter (Schmidli and Frei, 2005; Feldmann et al., 2012) for the observed record and the near future.

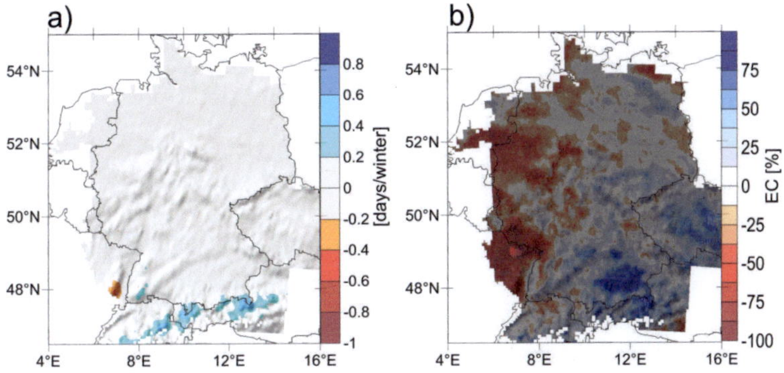

Figure 7.4: Ensemble mean climate change signal of compound cold and wet extreme days in winter (DJF) between 1971-2000 and 2021-2050. a) mean number of days/winter, b) ensemble consistency of a, grid points where the change signal is not significant are shaded.

7.3 Added value of high resolution

As minimum and maximum temperature are not available as daily output variable for most of the simulations, they were calculated as daily minimum and maximum values of the hourly mean temperature for the 7 km ensemble. For the 50 km ensemble, however, the temperature is only available every six hours. A comparison of both resolutions with observations would be possible by using a 6-hour mean of the observational and the 7 km data.

This would, however, smooth out all temperature extremes and not really fit the purpose of the comparison. Therefore, this chapter only compares the added value of precipitation extremes, namely the number of dry days in summer and the number of heavy precipitation extremes in winter. For better comparison, the uncorrected model precipitation values were used and compared to the HYRAS precipitation due to the known problems of E-Obs with heavy precipitation. Fig. 7.5 depicts the root mean square error and the correlation between the 7 km and 50 km ensemble and the HYRAS observations for dry days including all grid points of the HYRAS domain. The measures were calculated for each ensemble member separately and are displayed as box plots. As already noted in chapters 4 and 5, the COSMO-CLM has a drizzle problem and especially the 7 km ensemble greatly underestimates the number of dry days. At coarser resolution, this problem is slightly smaller but the difference is not significant. The correlations are fairly high for ensembles of both resolutions but the difference between the values for both ensembles are again not statistically significant.

When comparing RMSE and the correlation for wet extremes in winter (Fig. 7.6), the 7 km ensemble shows a clear added value for the spatial correlation. The RMSE shows a higher spread than the coarser resolved 50 km ensemble but the difference between the resolutions is not significant (note: the reason that the absolute value of the RMSE is lower for wet and dry extremes than for hot and dry extremes is also due to the fact that a large portion of the HYRAS domain only show very small numbers of wet days per winter, whereas the number of dry days is around 2 orders of magnitude higher). For elevations below 1000 m, the 7 km ensemble shows a significantly lower RMSE, but at higher elevation the RMSE is higher (not shown).

Figure 7.5: Root mean square error (RMSE) and spatial correlation for the mean number of dry days in summer (JJA) of the 7 km and 50 km ensembles with respect to E-Obs observations for the HYRAS domain (see Fig. 4.1). Gray bars show the RMSE/correlation of the ensemble mean.

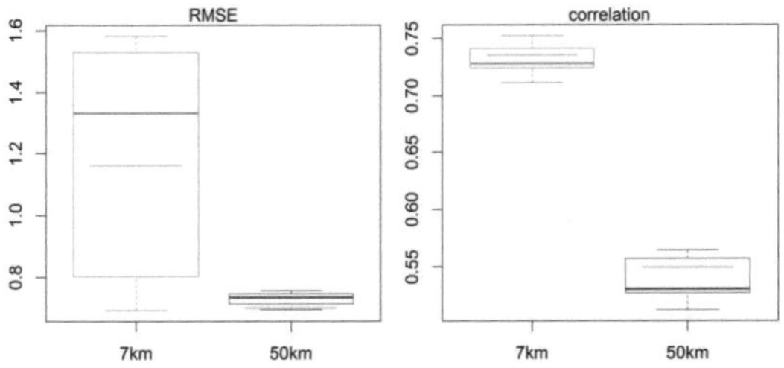

Figure 7.6: Root mean square error (RMSE) and correlation for the mean number of wet days in winter (DJF) of the 7 km and 50 km ensembles with respect to E-Obs observations for the HYRAS domain (see Fig. 4.1). Gray bars show the RMSE/correlation of the ensemble mean.

7.4 Short summary

In this chapter, compound extremes were defined as combined threshold exceedance of daily maximum temperature and precipitation sums (type 1 extremes), namely hot and dry extremes ($T_{max} > 30°$ and Prec < 1mm) in summer (JJA) and cold and wet extremes ($T_{min} < 0$ ° and Prec > 25mm) in winter (DJF). For the analysis, bias-corrected model data (see Chapter 5) was used.

Validation (1971-2000):

The number of compound hot and dry days in the reference period are between 0 and 11.1 days per summer in the investigation area with the highest number in the Rhine Valley and parts of Brandenburg. The ensemble mean is able to represent the statistics of the observations very well except for a small positive bias in the southern part of the investigation area. Compound cold and wet extremes mostly occur in the mountainous regions (Alps, Vosges and Black Forest), with a maximum of 5.4 days per winter. In the rest of the investigation area there are less than 0.5 days per winter. Including the uncertainties in observational datasets, the ensemble is able to represents the number of cold and wet extremes for the reference period well.

Climate change signal (1971-2000 vs 2021-2050):

The 7 km ensemble predicts a robust increase of compound hot and dry episodes in summer, averaged over the whole investigation area, the relative change is higher than 100%. Especially regions which already exhibit a high number of compound extreme days, such as the Rhine Valley and parts of Eastern Germany, are likely to experience the strongest absolute changes (up to 7.5 days per year) and a shift to more frequent and longer episodes. Near future changes of compound cold and wet extremes in winter are very small and mostly not significant. Only small parts of the eastern Alps show an increase of about 20%.

Added value of higher resolution (50 km vs 7 km):

A comparison of precipitation extremes for both the 7 km and 50 km ensembles results in added value in the spatial distribution of heavy precipitation by the 7 km ensemble.

8 Compound extreme events with the effective heat-/drought index: type 2 extremes

In contrast to the absolute compound extremes of the last one (Chapter 7), this chapter focuses on relative compound extreme events. These were analyzed by using the effective drought index (EDI; Byun and Wilhite, 1999) and an analogous measure defined for temperature. Relative compound extremes are defined as days where these indices show high deviations from the local mean conditions. Therefore, these extremes are not necessarily extreme in the sense of record breaking events. Nevertheless these relative extremes are of interest for different reasons. For one, they play an important role for agriculture and forestry, for example when regarding adaptation of seeding times of species, and they can affect the ecosystem. Furthermore, they are a measure of the variability of the climate system and the variability is also important when considering extreme events (e.g. Katz and Brown, 1992; Schär et al., 2004).
The extremes of this chapter are further referred to as type 2 extremes or relative compound extremes (different to the absolute extremes in the previous chapter). For validation, the HYRAS dataset (see Section 4.1.1), bilinearly interpolated to the 7 km model grid, was used.

8.1 The effective drought/heat index (EDI/EHI)

The effective drought index (EDI) was proposed by Byun and Wilhite (1999) and describes extremes as deviations from the climatological mean state.

As such, it is a measure of the variability of the climate system. A special feature of this index is the use of effective precipitation (hence the name), explained below, which takes the memory effect of the soil into account. An analogous measure was defined for temperature, called the effective heat index (EHI). Compound events are defined by both EDI and EHI exceeding a certain threshold on a given day.

Effective precipitation (hereafter referred to as EP) and effective temperature (ET) for a given day are calculated as described by equation 8.1 with EX = EP or ET. They are the weighted summation of the preceding ds days, thus taking into account the memory effects of soil and atmosphere. EP correlates highly with soil moisture which is especially important when considering droughts. The value of ds is different for temperature and precipitation. For the latter, the value suggested by Byun and Wilhite (1999) is used ($ds_{EP} = 365$). For the effective temperature, ds was determined as the lag where the autocorrelation function equals 0.5. This was calculated for every grid point of all ensemble members separately and then averaged, leading to a value of ds = 49. To have EP and ET synchronous, the starting date was set to the 1.1. 1972 using the preceding 365 days to calculate EP and the preceding 49 days to calculate ET. The same was done for the projection period where the calculation starts on the 1.1.2022.

$$EX_d = \sum_{n=1}^{ds} \left(\frac{\sum_{m=1}^{n} X_{d-m}}{n} \right) \qquad [8.1]$$

From EP and ET, the indices EDI and EHI can be calculated as standard anomalies of the effective values according to equation 8.2 where (X,Y) = (T,H) or (X,Y) = (P,D). The climatological mean value $\overline{EX_d}$ is calculated as a running mean over 5 days for precipitation (as suggested in Byun and Wilhite, 1999) and 31 days for temperature.

$$EYI_d = \frac{EX_d - \overline{EX_d}}{\sigma\left(EX - \overline{EX}\right)_d} \qquad [8.2]$$

Besides taking into account the memory effect, EDI and EHI have further advantages. One is the removal of linear biases. As shown in Chapters 4 and 5, the COSMO-CLM data is subject to a bias of varying magnitude. By using standardized anomalies, no bias correction needs to be applied, at least for linear biases. A further advantage of these indices is that they are symmetric and can thus be used for wet/dry and hot/cold extremes, respectively, and all combinations.

Since the aim of using this method is to capture deviations from the local mean state, detrended temperature time series are used for the calculation of EP to avoid a trend in the calculated EHI time series. A positive trend could possibly lead to stronger negative deviations at the beginning and stronger positive deviations at the end of each time period and prevent distinction between linear changes and the, in this case more interesting, changes in variability.

In this work, an extreme value of EDI/EHI is defined as a value greater than $\pm$ 1.5 which corresponds to a value greater than 1.5 times the standard deviation (for a normal distribution this would correspond to the 93th quantile, but note that the values are most likely not normally distributed). An exemplary time series for EDI and EHI is shown in Fig. 8.1. The "normal" range is marked by a gray box, all values lying outside of this box are extreme. Relative compound extremes are defined as both EDI and EHI exceeding $\pm$ 1.5, as for example the case of the yellow box.

The following sections show the results of the analysis of compound hot and dry extremes (EHI > 1.5 and EDI < -1.5) in summer (JJA) and compound cold and wet extremes (EHI < -1.5 and EDI > 1.5) in winter (DJF). Besides the number of extreme days, the mean number of episodes and the mean episode length are calculated. In contrast to Chapter 7, one extreme day is already considered an episode. This is justifiable, since due to the memory effect, a relative extreme with this method already implies extreme or nearly extreme conditions on the preceding days. The results are validated against HYRAS observations before a change signal is deduced.

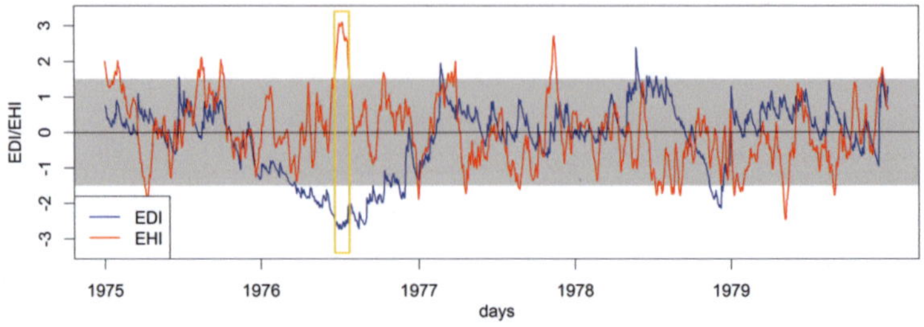

Figure 8.1: Exemplary EDI and EHI time series. All values outside of the gray shaded area show deviations greater than 1.5 σ and are considered extreme. The yellow box marks a compound extreme, where both EHI and EDI are greater/smaller than $\pm$ 1.5.

In order to assess the robustness of this change signal, the ensemble consistency (see Section 2.3) and the significance (see Section A) are calculated. In the chapters before, in most cases a high ensemble consistency implied significance and vice versa. For example, changes in mean temperature and temperature extremes are mostly significant and all members agree on the sign of the change signal due to the temperature trend which all models show. This relation between ensemble consistency and significance is not always valid, especially when looking at change signals which are derived from detrended time series, where the trend as strongest change signal is removed. This is exemplarily shown in Fig. 8.2. The boxplots represent the number of compound extreme hot and dry days of the eight ensemble members calculated from the EDI and EHI time series for two selected gridpoints. The boxes mark the interquartile range, whiskers the minimum/maximum, the black line the ensemble median and the gray line the ensemble mean. The p-value for the changes of grid point 1 (shown on the left side) is 0.28 (corresponding to a singificance level of 28%), the change is not significant according to the often used 5% significance level (p-value=0.05). The ensemble consistency in this case, however, is 100%, all members show a positive change greater than 10%. The change signal for the second grid point is significant

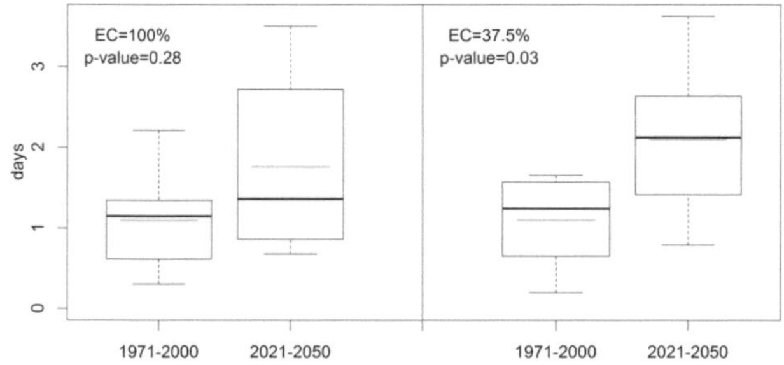

Figure 8.2: Exemplary comparison of ensemble consistency and significance of the change signal for compound hot and dry extremes for two different gridpoints of the HYRAS domain (see text).

(with a p-value of 0.03) but the ensemble consistency is only 37.5%. Five ensemble members show a positive change, two a negative change and one a change which is smaller than the threshold of 10%. Although the significance is an important measure in this case, it might be misleading and it is perhaps better to follow the thoughts of von Storch and Zwiers (2013) who propose to use "*a simple descriptive approach for characterizing the information in an ensemble of scenarios*" instead. They argue that the fundamental assumptions for building a null hypothesis are often not true when looking at climate ensembles. In this work, both measures are used.

8.2 Hot and dry extremes in summer

8.2.1 Validation

The statistical parameters of compound relative hot and dry extremes in summer (JJA) are shown in Fig. 8.3. This graphic also contains information about whether or not the statistical parameters of the observations lie within the ensemble spread (mean $\pm$ standard deviation). Gridpoints for which this is not the case are shaded.

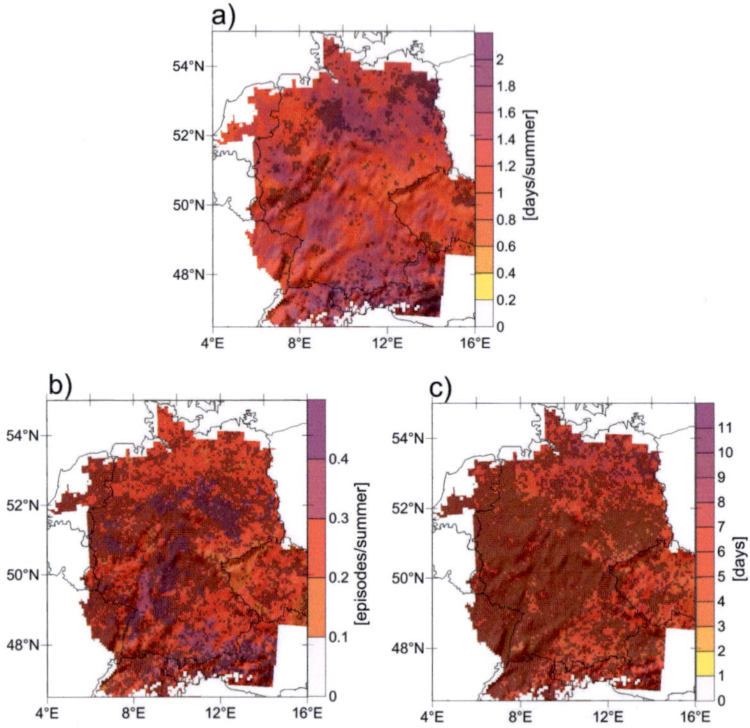

Figure 8.3: Ensemble mean of relative compound hot and dry extremes (EHI > 1.5 and EDI < -1.5) in summer (JJA) in the reference period (1971-2000). a) number of days/summer, b) number of episodes/summer, c) mean episode length. Gridpoints where the statistical parameters of the HYRAS Observations lie outside of the ensemble spread are shaded in gray.

The number of relative compound extreme days per summer lies between $\approx$ 0.5 and 2, the mean value for the HYRAS-domain is 1.2. In the areal mean, less than 2% of summer days are extreme. For comparison, the mean number of univariate extreme days per summer is 7.1 (hot days) and 4.7 (dry days) respectively, which amounts to about 8 %/5% of the total number of summer days and is still fairly extreme. The highest values of type 2 compound hot and dry extreme days are in the Rhine Valley (which also shows the highest number of absolute extremes, see Chapter 7), but also in the Alps and in northern Germany, south of Hamburg. These are regions where the number of absolute extremes is not extremely high or non-existent. For most gridpoints (82% of the HYRAS domain), the number of relative compound hot and dry days of the observations lie within the ensemble spread and the 7 km ensemble is able to represent the statistical parameters calculated from observations. For univariate temperature extremes, this is valid for 99% of the gridpoints (not shown), thus, the deviating factor is due to precipitation index. The number of episodes is lower than one episode per year, and the mean length lies between 2.7 and 10.4 days. For these latter two statistical parameters, the values calculated from observations lie inside of the ensemble spread for 45% and 32%, respectively, of the gridpoints within the HYRAS domain (non shaded areas in Figs. 8.3 b and c). For the other gridpoints, the ensemble shows too litte episodes with a higher mean episode length. Although the models capture the number of compound extremes fairly well, they seem to have problems to correctly simulate when these extremes occur and with which temporal succession.

8.2.2 Climate change signal

Fig. 8.4 shows the climate change signal of relative compound hot and dry extremes in summer for the HYRAS domain. The number of extreme days increases for a large fraction of the model domain. Especially in the eastern part of the investigation area (eastern Germany, Czech Republic and bordering parts of Austria) the relative increase is greater than 100% (see Fig. 8.3). The ensemble consistency is high for regions with a high change signal. The change in number of episodes is negligibly small (below 0.1 episodes/summer) for most parts of the investigation area. For gridpoints that show the highest increase of relative compound extreme days (Czech Republic, parts of eastern and northern Germany), the number of compound extreme episodes also increases.

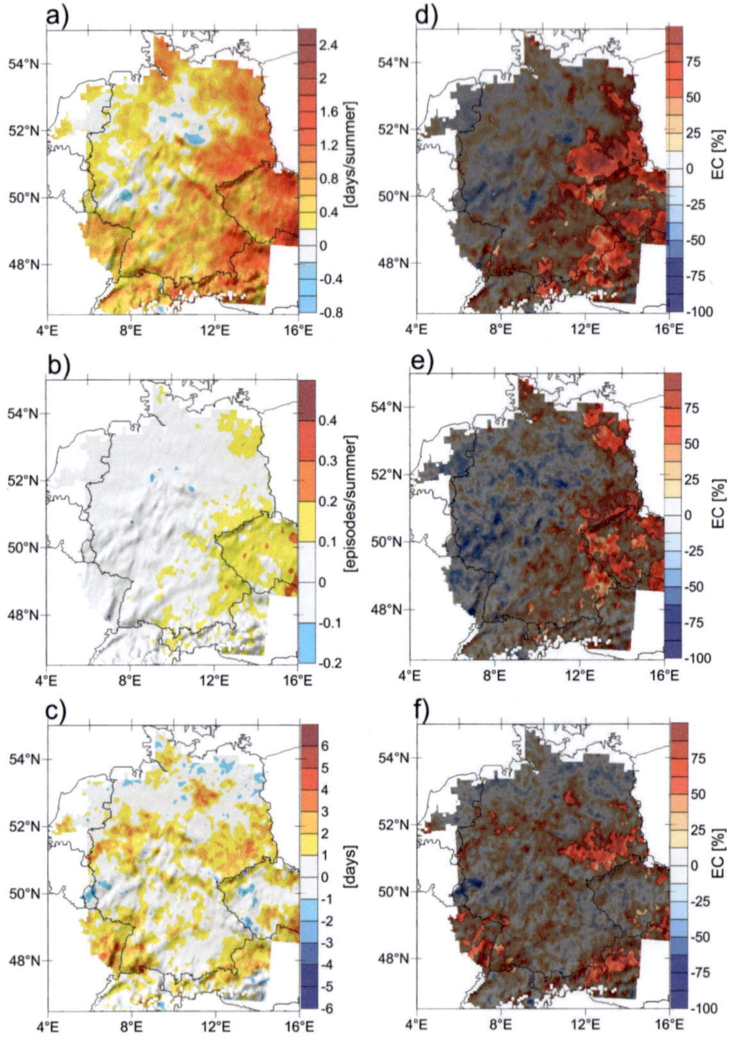

Figure 8.4: Ensemble mean climate change signal of relative compound hot and dry extremes (EHI > 1.5 and EDI < -1.5) in summer (JJA) between 1971-2000 and 2021-2050. Left side: ensemble mean, right side: ensemble consistency, gridpoints with non-significant changes at the 5% level (p-value = 0.05) are shaded in gray.
a) + d) mean number of days/summer, b) + e) mean number of episodes/summer, c) + f) mean episode length

The direction of change for the mean episode length depends on the region but for most gridpoints, the change signal does not show a high ensemble consistency and is not significant according to the 5% level (p-level=0.05). Exceptions are Austria and the parts of the Czech Republic as well as the bordering region in eastern Germany. Univariate dry days show a robust increase in most of the investigation area, which is in agreement with other studies (e.g. Dai, 2013) and increases the confidence of these findings.

8.3 Cold and wet extremes in winter

8.3.1 Validation

The statistical parameters for relative cold and wet extremes in winter are depicted in Fig. 8.5. The highest number of cold and wet days occurs in the eastern part of the model domain on the boarder between the Czech Republic and Austria ($\approx$ 0.86 days/winter) and the eastern Alps. The number of episodes is below 0.1 episodes per winter for all regions except the eastern Alps and areal mean of the mean episode length for the investigation area is 2.7 days. The statistical parameters of the observations only lie inside of the ensemble spread for less than half of the grid points (30%/ 48%/ 33%for cold and wet days/ episodes/ mean episode length). The ensemble underestimates the values for all three parameters. The statistical parameters for univariate cold and wet extremes are better represented by the ensemble (not shown), the percentage of gridpoints for which the observations lie within the ensemble spread is between 48 and 76%. Thus, the models do no correctly represent the combination of both extremes for a large fraction of the investigation area.

8.3.2 Climate change signal

The change signal of the statistical parameters of relative cold and wet days are shown in Fig. 8.6. The number of cold and wet days increases in small parts of northeastern and southeastern Germany (by up to 0.35 days per year) and decreases in parts of central Germany (by up to 0.3 days per year), the rest of the investigation area only shows changes smaller than 0.2 days per year (see Fig. 8.6).

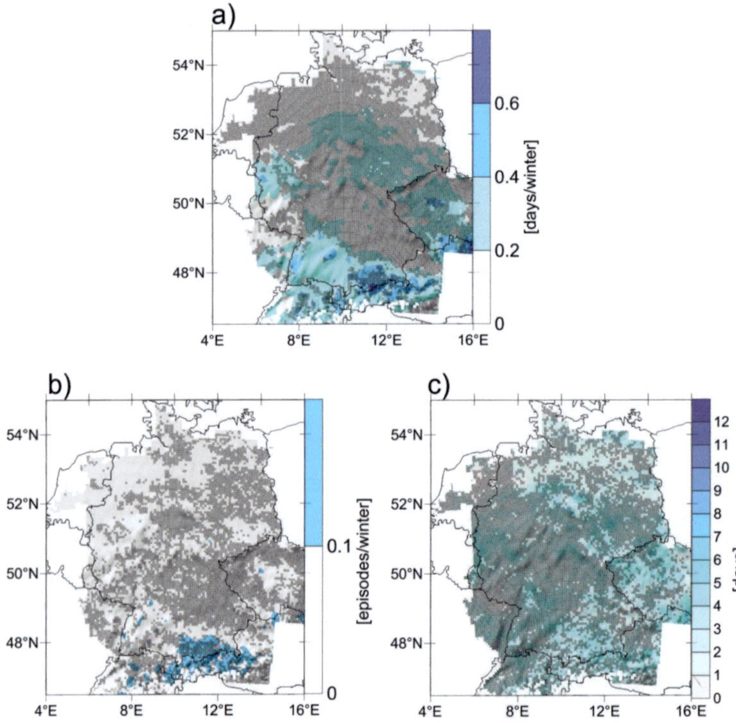

Figure 8.5: Ensemble mean of relative compound cold and wet extremes (EHI < -1.5 and EDI > 1.5) in winter (DJF) in the reference period (1971-2000). a) mean number of days/winter, b) mean number of episodes/winter, c) mean episode length. Gridpoints where the statistical parameters of the HYRAS Observations lie outside of the ensemble spread are shaded in gray.

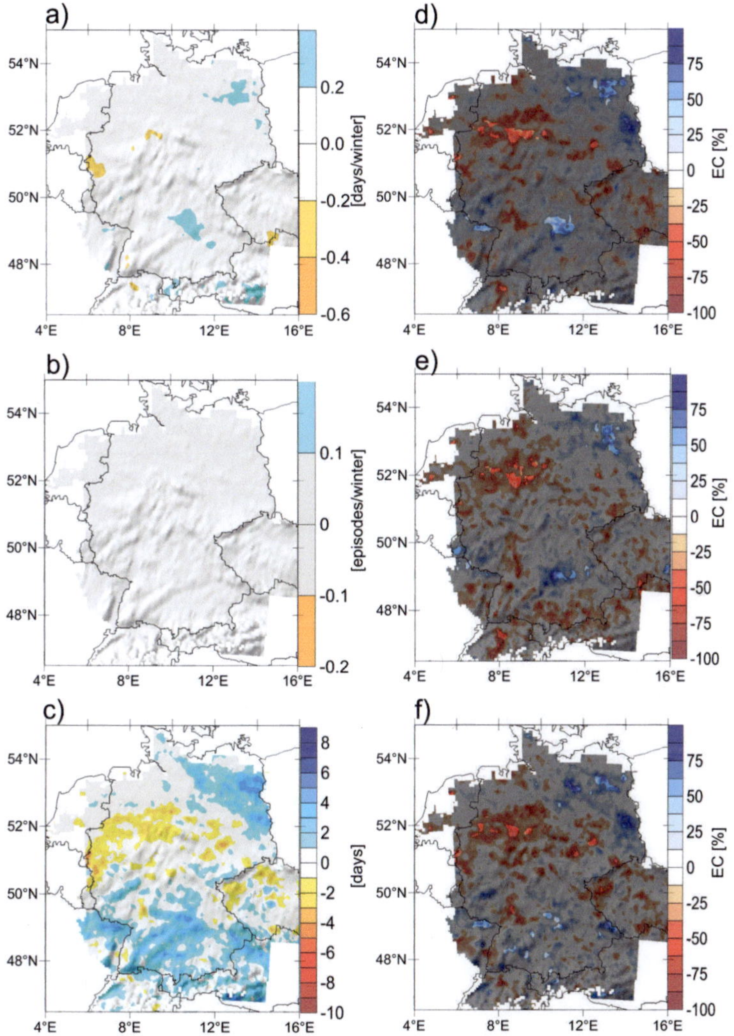

Figure 8.6: Ensemble mean climate change signal of relative compound cold and wet extremes (EHI < -1.5 and EDI > 1.5) in winter (DJF) between 1971-2000 and 2021-2050. Left side: ensemble mean, right side: ensemble consistency, gridpoints with non-significant changes at the 5% level (p-value = 0.05) are shaded in gray. a) + d) mean number of days/winter, b) + e) mean number of episodes/winter, c) + f) mean episode length

The change in number of episodes is negligibly small and the mean episode length shows a decrease in the southern and northeastern parts of the domain and an increase in the western part and parts of the Czech Republic. However, the ensemble consistency of the change signal for all three statistical parameters is only high for small regions, only less than 20% of the gridpoints show a consistency higher than 50%, and the results are significant for less than 5% of the investigation domain.

8.4 Added value of high resolution

For the analysis of added value of the 7 km resolution ensemble, the statistical parameters for compound hot and dry/cold and wet extremes are compared for the reference period. Precipitation and detrended temperature data of the 50 km ensemble were interpolated to the 7 km model grid by bilinear interpolation prior to the calculation of the indices. As a measure of added value, the root mean square error (RMSE) including all grid points of the HYRAS domain and the spatial correlation of the respective statistical parameter (number of days, number of episodes and mean episode length) with respect to HYRAS observations were calculated for each ensemble member separately.

The results are shown as box plots where the box marks the interquartile range of the ensemble and the whiskers the minimum and maximum RMSE and correlation of the eight ensemble members. The colored bars mark the median and the gray bars the values for the ensemble mean. Box plots for 7 km and 50 km ensemble are pictured next to each other for comparison. Fig. 8.7 shows the RMSE for hot and dry extremes in summer. While the number of compound extreme days seems to be slightly better represented by the 7 km ensemble and the mean episode length slightly worse, there are no significant differences between the ensembles of different resolutions. The same can be said for the cold and wet extremes in winter (Fig. 8.8). When looking at relative extremes, the second nesting stage (7 km) does not seem to differ much from the first nesting stage at 50 km. As these are related to the variability of the extremes, this seems to be mostly governed by large scale circulations.

8.5 Short summary

The relative (type 2) compound extremes analyzed in this chapter refer to extremes as deviations from the local mean state (of the respective time period) and are a measure of the variability of the climate system. Temperature trends were removed prior to the analysis.

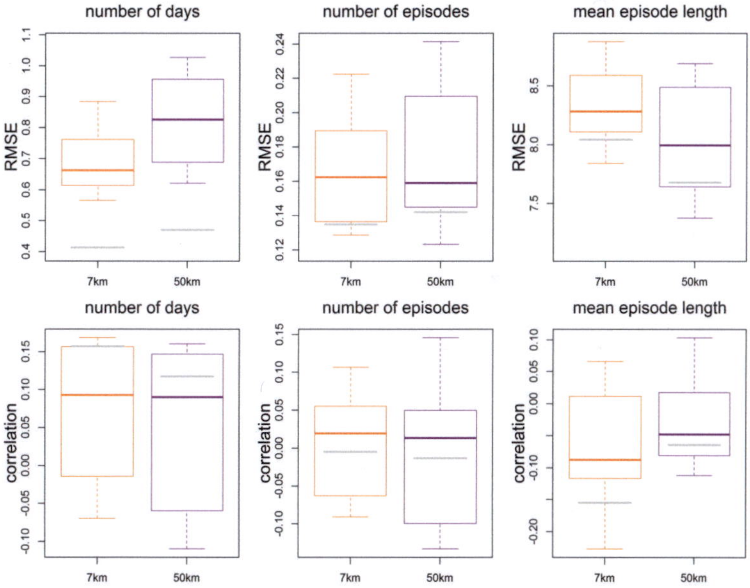

Figure 8.7: Root mean square error (RMSE, top row) and spatial correlation (bottom row) for relative compound hot and dry extremes in summer (JJA) in the reference period (1971-2000) over the HYRAS domain (see Fig. 4.1). Boxplots for 7 km and 50 km ensemble (see text).

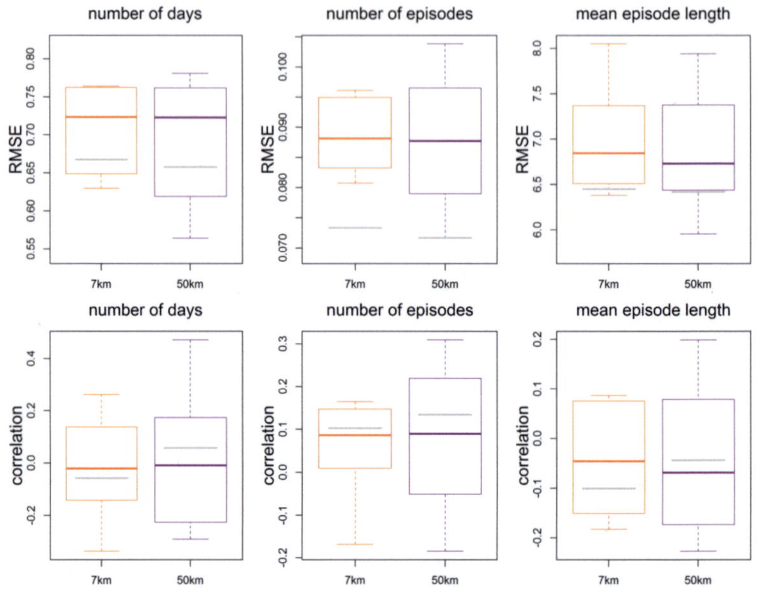

Figure 8.8: Root mean square error (RMSE, top row) and spatial correlation (bottom row) for relative compound cold and wet extremes in winter (DJF of 1971-2000) of the 7km and 50km ensemble members with respect to HYRAS observations for the HYRAS domain in the reference period (1971-2000) (see text). Gray bars show the RMSE/correlation of the ensemble mean.

Validation (1971-2000):

The number of compound hot and dry extreme days lies between 0.5 and 2.0 days per year in the investigation area, the highest number of days occur in the Rhine Valley, the Alps and northern Germany, south of Hamburg. The statistical parameters of the ensemble match those of the observations fairly well for the number of compound extreme days but the number of episodes are under-, the mean episode length overestimated by the ensemble (percentage of gridpoints where the observational value lies within ensemble spread = 85%/45%/32% for number of days/number of episodes/mean episode length). Type 2 cold and wet extremes in winter (DJF) occur mostly in the border between the Czech Republic and Austria and the eastern Alps. Compared to the HYRAS observations, the ensemble represents their number well in the southern part of the investigation area, in the rest the statistical parameters of the observations lie outside of the ensemble spread.

Climate change signal (1971-2000 vs 2021-2050):
Relative hot and dry compound extremes in summer (JJA) are likely to increase in most of the investigation area, especially in eastern Germany and the Czech Republic there are some regions where the 7 km ensemble predicts a significant increase of $\approx 100\ \%$ with most model members agreeing on this change. For cold and wet extremes, there are almost no robust changes in the investigation area.

Added value of higher resolution (50 km vs 7 km):
Significant added value for higher resolution was not found for the analyzed relative compound extreme events.

9 Compound extreme events with the Markov Chain method: type 3 extremes

Most methods used for the analysis of extreme or compound extreme events focus on the absolute number of events, their return periods or the variability (as for example the results presented in the two preceding chapters). The method presented in this chapter is a new approach for the analysis of compound extreme events which concentrates more on how they occur – on the temporal succession and interplay of different univariate extremes. This will be referred to as the dynamical behavior of compound extreme events or type 3 extremes. The method yields supplementary information to the existing methods and enables the analysis of an aspect of current climate and climate change which is usually neglected. If, for example, two regions show a similar number of extreme events but different dynamical behavior (i.e. in one region there are many short extreme episodes, in the other there are few but long ones) this has a huge impact on how these extremes affect society. In addition to this temporal succession, the method also yields information about the predictability of the system with regard to the compound extreme events. Considering changes of extremes, the method can be used to answer the question of whether or not changes in the mean or variability also induce changes in the dynamical behavior with respect to the new "normal" state with changed mean and variability and whether or not the predictability of compound extreme events changes.

9.1 Markov chain analysis

The method presented here for the analysis of the dynamical behavior of compound extreme events is based on the concept of Markov chains. Descriptors which characterize this dynamical behavior are calculated from the time series of atmospheric variables which are reduced to a symbolic sequence of extreme and non-extreme regimes beforehand. This sequence can be described as a Markov chain. The method is an adaptation of work by Mieruch et al. (2010), who first introduced it to climate science. They used it for climate classification and a comparative study of two regions based on temperature and water vapor data. Before that, it has been used in biology by Hill et al. (2004) to describe dynamics of succession of a rocky subtidal community.
The following sections give a brief review of Markov chains before introducing the descriptors used for this work and the application of the method to climate data is explained.

9.1.1 Markov chains

Markov chains are a class of time and state discrete models, or stochastic processes, used to represent time series of discrete variables (e.g., Norris, 1998; Wilks, 2011). They consist of m different states (m-state Markov chain) of a model system which are "mutually exclusive and collectively exhaustive" (Wilks, 2011) and, thus, make up the sample space of the random variable considered. For each discrete time step, the system can either stay in the state it is already in or change to another state. Conditional probabilities, which govern the behavior of the Markov chain for these transitions, can be calculated. The simplest form of Markov chain is a first order Markov chain. It fulfills the Markov Property,

$$P(x_t|x_{t-1},x_{t-2},\dots,x_{t-n}) = P(x_t|x_{t-1}), \qquad [9.1]$$

where the present state x_t is only dependent on the preceding state x_{t-1}. An m-state Markov chain allows $m \times m$ different transitions for which conditional transition probabilities can be organized in a transition probability matrix **P** of the order $m \times m$ (m= number of discrete states of the Markov chain). The entries for the different transitions of state $j = \{1,..,m\}$ and time t to state $i = \{1,..,m\}$ at time $t+1$ of **P** can be estimated as follows:

$$\hat{p}_{ij} = \frac{n_{ij}}{\sum_i n_{ij}}, \quad [9.2]$$

where n_{ij} is the total number of transitions from state j to state i. Note that the entries of each column $\sum_i \hat{p}_{ij}$ must equal 1 since every transition must be into one of the other states. In this work, homogeneous first order Markov chains are used for which the transition probability matrix **P** is time independent. Additionally none of the entries of the transition probability matrix should be equal to zero. To test for stationarity, the stationary distribution $\boldsymbol{\pi}$ and the empirical distribution $\hat{\pi}$ of the Markov chain need to be identical. A stationary distribution is a vector that fulfills the following equation:

$$\boldsymbol{\pi} = \mathbf{P}\boldsymbol{\pi}. \quad [9.3]$$

The stationary distribution $\boldsymbol{\pi}$ can therefore be obtained by solving the eigenvalue problem of equation 9.3. $\boldsymbol{\pi}$ is the eigenvector associated with the dominant eigenvalue of 1. The empirical distribution can be calculated by counting the different states of the Markov chain:

$$\hat{\pi}_j = \frac{n_j}{\sum_j n_j}. \quad [9.4]$$

9.1.2 Markov descriptors

To characterize the dynamical behavior of compound extreme events, descriptors can be calculated from the estimated transition probability matrix

P of the Markov chain. Following Mieruch et al. (2010), this work focuses on only three of the descriptors mentioned in Hill et al. (2004): persistence, recurrence time and entropy. These descriptors can either be calculated for the whole sample space or for single states. In this work, the focus lies on the single state definition of the descriptors since this is the one which has been used for the analysis of compound extreme events.

Persistence:

The persistence P_j is a measure of duration of the compound extreme event. It is calculated as the diagonal entry of the transition probability matrix **P**

$$P_j = \hat{p}_{jj} \tag{9.5}$$

and gives the probability that the system will stay in the same state in the next time step. The persistence of the extreme state thus gives the probability that the system will reside in this extreme state. The theoretical limits are 0 (the system will always change to another state and there will be no two consecutive extreme days in a row) and 1 (if the system is in an extreme state it will stay there, all extreme states follow each other).

Recurrence time:

The recurrence time R_j is the mean time the system needs to get back to the extreme state:

$$R_j = \frac{1 - \hat{\pi}_j}{(1 - \hat{p}_{jj})\,\hat{\pi}_j}. \tag{9.6}$$

It is connected to the persistence $\hat{p}_{jj}$ as well as to the total number of extreme events through the stationary distribution $\hat{\pi}_j$. The theoretical limits are 0 (the system always stays in the same state, corresponding to a persistence of 1) and ∞ (the system never comes back to the extreme state, note: this does not correspond to a persistence of 0).

Entropy:
The entropy $H(p_j)$is a measure based on the fundamental works on information theory by Shannon (1948) and is an inverse measure of the predictability of the Markov chain. The conditional probabilities of transitions from the state of interest j to all possible states i (including state j) are included in the calculation. The normalized single state entropy can be calculated by:

$$H(p_j) = -\frac{\sum_i \hat{p}_{ij} \log \hat{p}_{ij}}{log\left(\frac{1}{m}\right)}. \quad [9.7]$$

Therefore, unlike persistence and recurrence time, which depend only on the compound extreme state of interest, the entropy additionally depends on the transitions to the other states and is therefore more susceptible to the way these states are chosen. The theoretical limits of the entropy are 0 (which means that the system is deterministic and the next state is always known when in an extreme state) and 1 (the system is random and the next state cannot be predicted). The entropy can be used to identify and characterize complex dynamics like deterministic chaos, which is not possible with simple linear methods.
The actual, empirical limits of the descriptors are smaller and will be discussed at the end of the next section.

9.1.3 Application to climate data

These descriptors can be applied to climate data for different purposes. Since the correct representation of the dynamical behavior of compound extreme events in models is a requirement for deriving their climate change signal, comparison and validation of different climate models, or climate models with different configuration and at different resolutions are one possible application of the method. Furthermore, regional differences can be assessed. A further interesting application is the comparison of different observational datasets. These are usually validated and compared with respect to their mean values, variability and extremes as well as cross-correlation between different variables, but the successional dynamics are usually not part of the validation. In addition, the assessment of a climate change signal from climate model data is of interest. In this thesis, the underlying hypothesis for this is that a linear trend induced by external forcing, as for example the temperature trend, also induces a change in the internal dynamical behavior of the climate system with respect to the extreme events. While changes in trends and variability have been thoroughly studied, analyses as with this method are rare and the results cannot always be intuitively predicted (unlike, e.g., as the connection between a positive temperature trend and the increase of absolute hot days).

The application of this method to climate data can be divided into the following four steps:

1. Preprocessing of data: The main focus of this method is the dynamical behavior of the climate system. Therefore, daily anomalies are used and all linear trends and annual cycles removed from the data. In this work the focus lies on compound temperature and precipitation extremes, and anomalies of detrended temperature time series are used. For looking at dry conditions, the EDI (see Chapter 8) is used as it is highly correlated to soil moisture and with that a better measure for describing drought than precipitation itself. In this chapter, the EDI for the near future is calculated with the standard deviation σ of the near future, different to Chapter 8. For heavy precipitation events, seasonal anomalies of precipitation are used.

   Although the long term trend is removed from the data, the hypothesis is that changes between different time periods in the descriptors are nonlinear effects of this linear forcing.

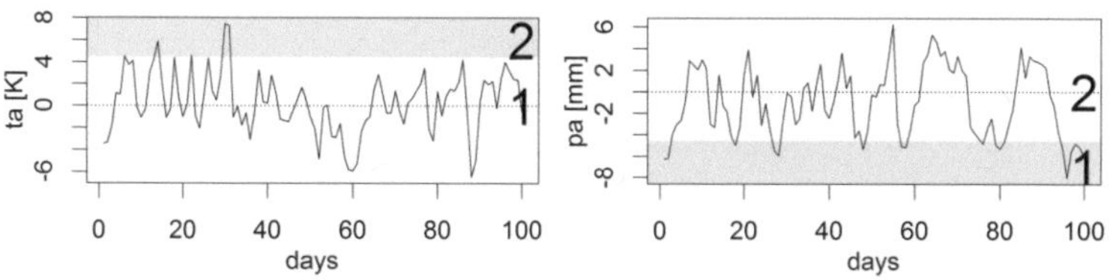

Figure 9.1: Partitioning of temperature anomalies (ta) and precipitation anomalies (pa) into extreme (gray areas) and non extreme (white areas) states.

2. Construction of a Markov chain: To construct a Markov chain from the anomalies of atmospheric variables, in a first step the univariate time series are reduced to a symbolic sequence of extreme and non-extreme regimes. This is exemplarily shown in Fig. 9.1. The concept of this coarse grained representation of data comes from symbolic dynamics and is referred to as partitioning (Freund, 1996). These 2-state symbolic sequences, or 2-state Markov chains, are then combined to a multivariate symbolic sequence of $m = 2^{\nu}$ different states (ν number of variables). For the case of compound temperature and precipitation extremes, 4 states are possible, listed in Table 9.1. Because the method is sensitive to the absolute number of extreme events, percentiles are used to partition the data. This way, the number of univariate states remains the same for all ensemble members, different grid-points/regions and different time periods thus enabling a regional comparison of the descriptors or the analysis of a climate change signal.

   The thresholds used for the two extremes analyzed in this work, namely hot and dry extremes in summer and cold and wet extremes in winter, are summarized in Table 9.2. It needs to be kept in mind that only the number of univariate extreme events is kept constant. The temporal correlation of these univariate events can be different for another region or time period and thereby the absolute number of compound extreme events. For example the total number of compound hot and dry extremes is higher than the number of cold and wet extremes although the number of univariate extremes is the same in both cases. Furthermore, different regions show a different number of compound extreme events (compare, e.g., Figs. 9.4 a and 9.10 a). The thresholds were chosen to obtain a balance between meeting the requirements of stationarity (non-zero entries of the column j of the transition probability matrix) and still being in an extreme state. When calculating a climate change signal between the descriptors of the reference period and the near future, the extreme state for both time periods are in relation to the mean climate for that respective time period. In addition to a change of the mean values, the threshold with respect to the changed mean can also change. This is equivalent to a change in variability. The changes of threshold values between the two time periods are shown in Figs 9.2 and 9.3.

Table 9.1: Partitioning for compound hot and dry and compound cold and wet extremes

state no	partitions	symbols T,P	compound hot and dry extremes	compound cold and wet extremes
1	$T < th_{t1}, P < th_{p1}$	1,1	normal and dry	cold and normal
2	$T < th_{t1}, P \geq th_{p1}$	1,2	normal state	cold and wet
3	$T \geq th_{t2}, P < th_{p2}$	2,1	hot and dry	normal state
4	$T \geq th_{t2}, P \geq th_{p2}$	2,2	hot and normal	normal and wet

Table 9.2: Thresholds for daily temperature and precipitation extremes.

hot extreme	T anomaly > 90th quantile
cold extreme	T anomaly < 10th quantile
dry extreme	EDI < 25th quantile
wet extreme	Seasonal precipitation anomaly > 75th quantile

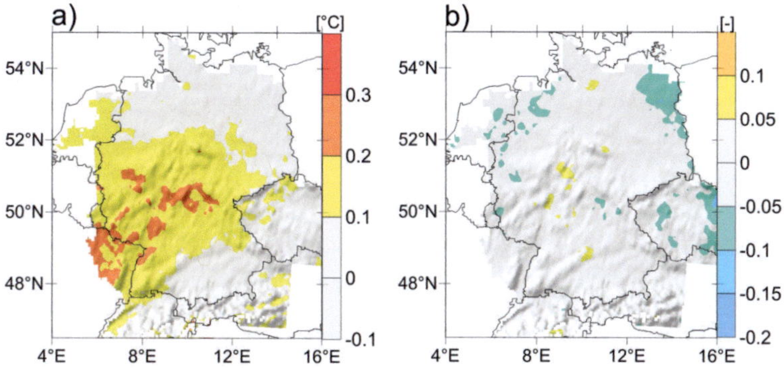

Figure 9.2: Changes in temperature and precipitation thresholds between summers (JJA) of 1971-2000 and 2021-2050: 90th percentile of temperature (left) and 25th percentile of the EDI (right).

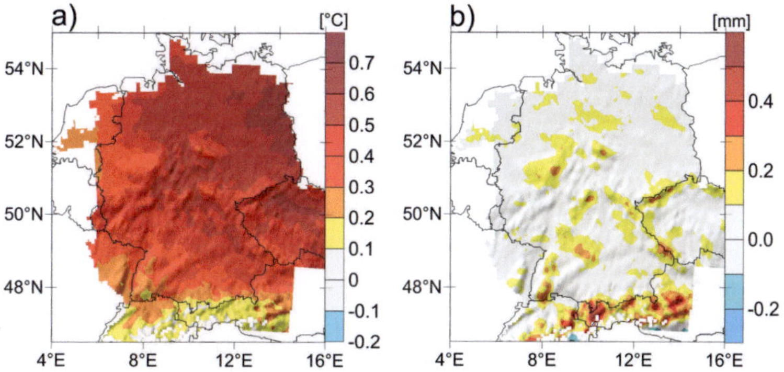

Figure 9.3: Changes in temperature and precipitation thresholds between winters (DJF) of 1971-2000 and 2021-2050: 10th percentile of temperature (left) and 75th percentile of precipitation (right).

The temperature threshold increases for both cold and hot extremes between the two time periods which means that the deviations are larger in the future and the magnitude of the variability increases. Changes in precipitation thresholds show an increase for winter high precipitation events and a decrease for summer dry events but only in some regions. These changes are all excluded from the analysis as they can be assessed with other methods (e.g. the ones described in Chapters 8 and 9). The state in the future with changed mean and variability is taken as the new "normal" state when calculating a climate change signal.

3. Calculation of transition probabilities: Transition probabilities are calculated as explained in Section 9.1.1. For the stationarity test, a deviation smaller than max ($\hat{\pi}$*0.1, 0.001) between the empirical distribution $\hat{\pi}$ and stationary distribution $\boldsymbol{\pi}$ is allowed. Additionally, the entries p_{ij}, with j = compound extreme state of interest, have to be non-zero.

4. Calculation of descriptors: Persistence, recurrence time and entropy are calculated according to Section 9.1.2.

A great advantage of this method is that it can in theory be applied to compound extremes of as many variables as wanted, although the computational efforts increase with the number of variables. Furthermore, all linear biases are removed due to the use of anomalies and the partitioning. To better understand the descriptors, their dependencies and limits and to get an idea on how to interpret their climate change signal, an exemplary case is explained below.

Understanding the descriptors:
Persistence only depends on the extreme state. It counts the number of transitions n_{jj} from one compound extreme state to another compound extreme state. The lowest persistence, 0, is reached when no two extreme states succeed each other, whereas the highest persistence is reached when all states are in a row. In this latter case, persistence is calculated by the maximum number of possible transitions (which is one less than the total number of extreme states) divided by the total number of extreme states: $(\sum i_e - 1)/\sum i_e$. With growing sample size, this equals nearly one (e.g. for 100 compound extreme

states it is 99/100 = 0.99). The actual limits of the persistence depend strongly on the type of the compound extreme considered. In the case of temperature and precipitation extremes, a persistence of $P_j \approx 1$ is nearly impossible under current climate conditions. If, for example, daily data of 30 summers are analyzed, this amounts to a total of 2700 days (states). With the partitioning used in this work (see Table 9.2), about 3% or 90 days are compound hot and dry extremes. A persistence of ≈ 1 in this case would mean that all of these compound hot and dry states occur consecutively. This would correspond to all days in one whole summer (JJA) of one year having stronger deviations from the mean state than any day in all other summers. In central Europe this is not very likely or would indicate an extreme shift in our climate as even the 2003 heat wave did not fulfill this criteria. For compound events including heavy precipitation, a persistence of 1 is even more unlikely as precipitation shows a higher temporal variability. The theoretical lower limit of 0 is possible for compound extremes which are few in number and have a high variability. It is more likely for events related to heavy precipitation than for temperature or drought as these variables exhibit a higher autocorrelation and are usually grouped in some way. However, for a persistence of 0 the stationarity criteria are not met. Since the limits depend strongly on the partitions chosen and the type of extreme considered, no general empirical limits can be given, but when comparing the descriptors the data should be chosen in the same way (e.g. same number of time steps, same partitions $\equiv$ same number of univariate extremes) and additional information (like the total number of compound extremes) has to be be consulted for the analysis. The recurrence time is connected to the persistence. A high persistence implies a long recurrence time since the mean time between the compound extreme events will be high. A low persistence implies a shorter recurrence time as the extremes occur more frequently. Additionally the recurrence time depends on the total number of compound extreme states, it is shorter for a high number of states and longer for a low number of states.

The descriptor which is perhaps hardest to intuitively understand is the entropy. It does not only depend on the extreme states but also on the transitions from the extreme state to other states. By this, it is also strongly dependent on the partitioning. If, as for example, in a two state Markov chain there is only one other possible state to change to, the entropy will generally be lower than if there are two, three or more possible other states. The entropy is lowest if there is one favored transition, therefore the lowest entropies come along with very high or very low persistences with most extreme state to non-extreme transitions into the same non-extreme state (no matter which one). In order to better understand this, some examples are shown for a Markov chain of which 100 states are compound extreme states, corresponding to partition 1 (the total number of states is irrelevant for this example). With these 100 extreme states, 100 transitions are possible. To calculate the entropy, only the column p_{i1} of the transition matrix **P**, holding the probabilities for transitions from the extreme state 1, is needed. Some examples (which not necessarily make sense for climate data but are helpful to understand the concept) are shown in Table 9.3. Multiplying the entries of p_{i1} by the number of states (100) gives the number of transitions between the respective states. The first entry of p_{i1},p_{11}, is the persistence of state 1 (compound extreme state). Example Table 9.3 a shows an equal distribution of transitions. No state is favored over the other and the entropy is 1, the system is completely random. Example Table 9.3 b1 shows the highest possible persistence in the case of 100 extreme states without violating the stability criteria (no non-zero entries, 1 transition of 100 possible transitions yields 0.01), and the entropy is 0.12. Case Table 9.3 b2 has the same entropy, but the persistence is now P_j=0.01. For the entropy, the order in the column does not matter, only the effective numbers. The entropy is small for very unequal distributions as these are more predictable. Thus, for larger sample sizes, smaller entropies are possible without violating the stability criteria. Cases Table 9.3 c and d show the upper and lower limit of the entropy with a persistence of P_j=0.4 and, thus, 40 extreme to extreme transitions (note: only for this given example). Example

Table 9.3 c is the most equally distributed, the remaining 60 transitions are equally divided between the three non-extreme states thus the entropy is high because it is not very predictable. Example Table 9.3 d is the most unequal distribution possible for 100 compound extreme states and a persistence of 0.4 and, thus, the easiest to predict (lowest possible entropy for P_j=0.4). The distribution of the other states plays a role inasmuch the number of transitions to the other states are of course also more probable if there is a higher total number of that state. The total number of compound extreme events changes the lower limit, since with more states, a more unequal distribution which is more predictable is possible, leading to a lower entropy.

The descriptors are calculated for each ensemble member separately. The time series of all grid points are partitioned separately and then the symbolic sequences merged over a running window of 3×3 grid points to increase the data length and smoothen the results. Descriptors are then assigned to the center grid point.

Table 9.3: Entries p_{i1} of the column of the transition probability matrix corresponding to transitions from the extreme state to the other states from a total of 100 extreme states ($p_{i1} * 100$ = number of transitions). Bold entry: persistence,H_j: entropy .

a)	b1)	b2)	c)	d)
$p_{i1} = \begin{pmatrix} \mathbf{.25} \\ .25 \\ .25 \\ .25 \end{pmatrix}$	$p_{i1} = \begin{pmatrix} \mathbf{.97} \\ .01 \\ .01 \\ .01 \end{pmatrix}$	$p_{i1} = \begin{pmatrix} \mathbf{.01} \\ .01 \\ .97 \\ .01 \end{pmatrix}$	$p_{i1} = \begin{pmatrix} \mathbf{.40} \\ .20 \\ .20 \\ .20 \end{pmatrix}$	$p_{i1} = \begin{pmatrix} \mathbf{.40} \\ .58 \\ .01 \\ .01 \end{pmatrix}$
H_j=1	$H_j = 0.12$	$H_j = 0.12$	H_j=0.96	H_j=0.56

9.2 Hot and dry extremes in summer

9.2.1 Validation

For most evaluations of observational datasets, mean values, trends and spatial correlation are compared, sometimes the occurrence of extreme events assessed but a comparison of dynamical aspects of the extremes is not a standard procedure. Therefore, the descriptors of the two observational datasets (E-Obs and HYRAS) are compared prior to the validation of the ensemble. Fig. 9.4 shows the total number of compound hot and dry extreme events (note these are relative extremes, partitioned as described in the previous section) in the reference period for the HYRAS dataset (left side) and by how much the E-Obs dataset differs (right side). The number of events is highest in the south and southwestern part of the model domain and along the coast in the northeast. The E-Obs dataset differs in some regions, mostly in the southern and southeastern part of the investigation area where it yields a higher number of events (blue colors in Fig. 9.4, right side). Reasons for these deviations might be the different density of stations and the method of interpolation.

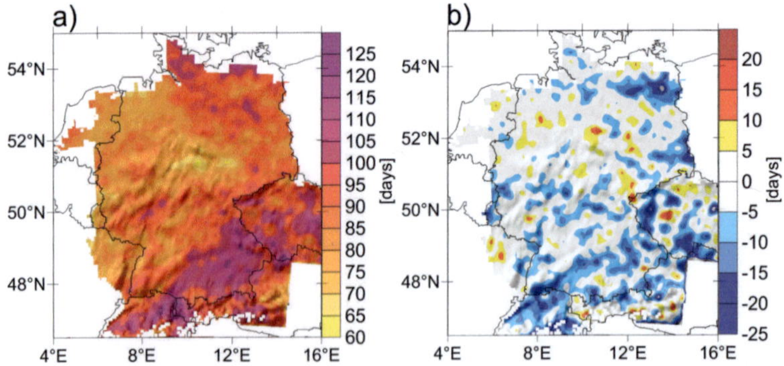

Figure 9.4: Number of compound hot and dry days in summer (JJA), 1971-2000. a) HYRAS, b) HYRAS-E-Obs.

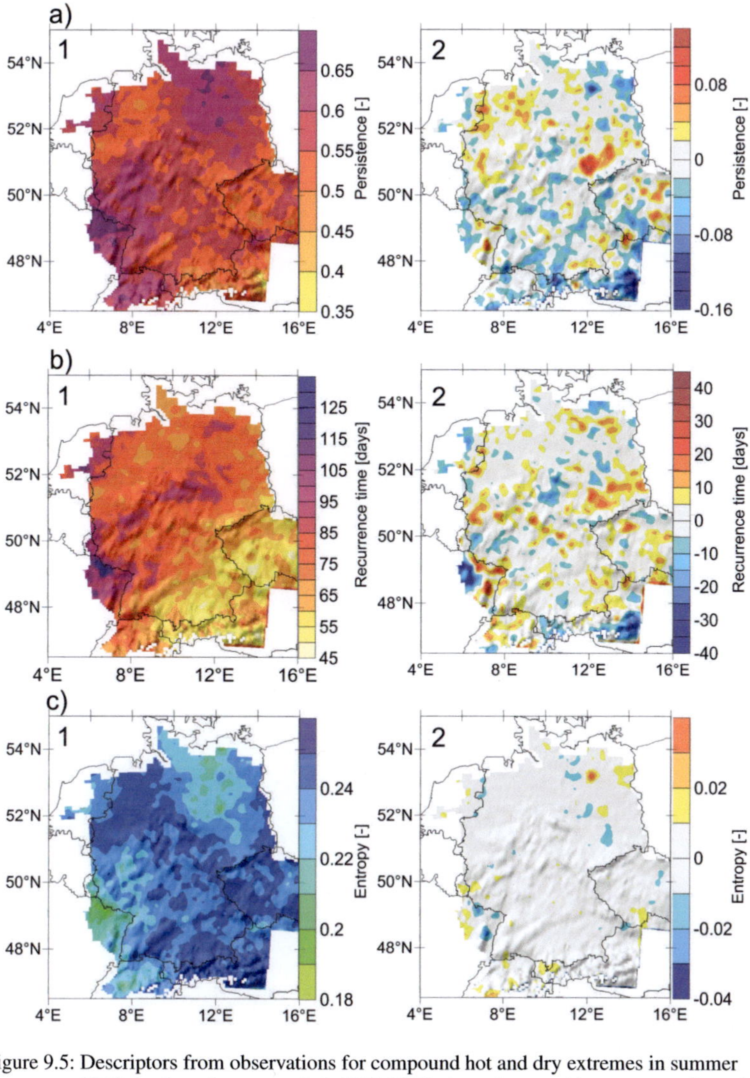

Figure 9.5: Descriptors from observations for compound hot and dry extremes in summer (JJA) for the reference period, 1971-2000. a) persistence, b) recurrence time, c) entropy. 1) HYRAS descriptors, 2) HYRAS-E-Obs

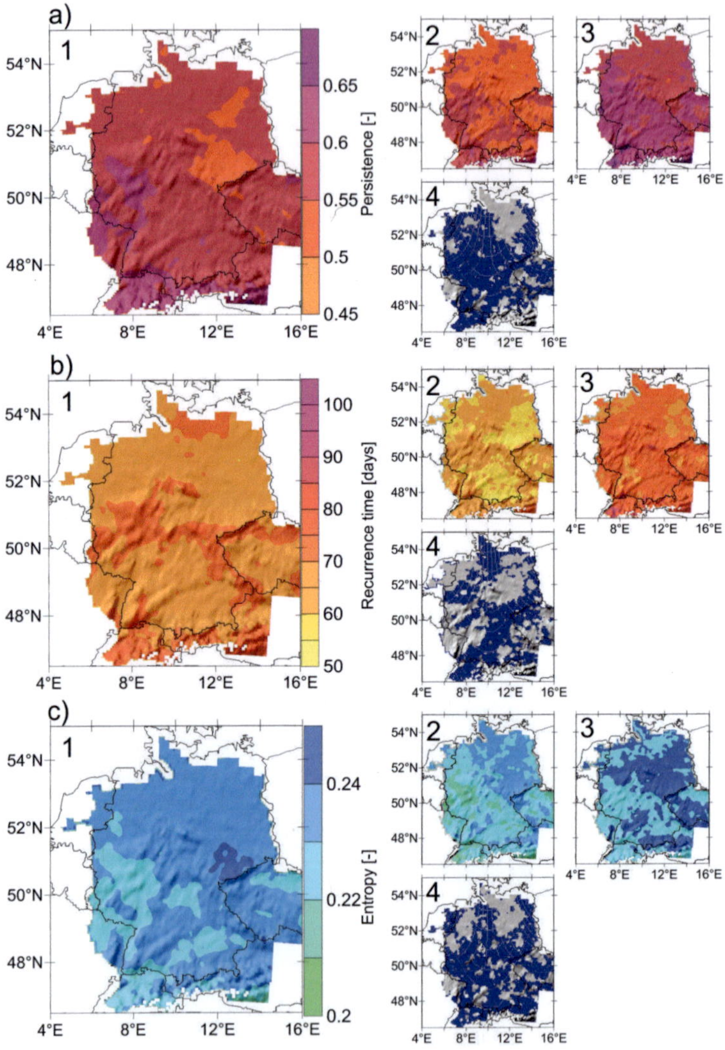

Figure 9.6: Descriptors from the 7 km ensemble for compound hot and dry extremes in summer (JJA) for the reference period, 1971-2000. a) persistence, b) recurrence time, c) entropy. 1: ensemble mean, 2: 1st quantile, 3: 2nd quantile, 4: grid points where HYRAS descriptors are within the ensemble spread (blue).

In Fig. 9.5 the descriptors for the two observational datasets are shown for 1971-2000, left the results for the HYRAS dataset and the deviation of E-Obs descriptors on the right. Regions with a high persistence and a low recurrence time show the most extreme behavior since they yield long and frequent episodes. The persistence for the investigation area lies between 0.37 and 0.71 with a mean of 0.57 for the HYRAS domain. This means that the probability for an extreme-extreme transition lies between 37 and 71%. The highest persistence is calculated in eastern France and northern Germany, the lowest ones in the eastern Alps. The E-Obs dataset shows strong deviations of the persistence, in some parts it is higher than the HYRAS persistence by 0.17, which is equivalent to the probability of an extreme-extreme transition being 17% more likely. The spatial correlation between the two persistences for the HYRAS domain is 77%.

The recurrence time is lowest in the southeastern part of the model domain, where the persistence is also comparatively low, but the events occur more frequently. This is also the region with the highest number of compound extreme events (see Fig. 9.4). The entropy lies between 0.19 and 0.25 and is lowest (highest predictability) in regions with high persistence. As discussed before, the entropy is lower if there is one favored transition and as such is correlated to high persistences. From the E-Obs dataset, compound extreme events are slightly less predictable (higher entropy) in most of the HYRAS domain. The spatial correlation between of E-Obs and HYRAS is 85% for the recurrence time and 80% for the entropy.

Although the two datasets exhibit a relatively high correlation they do show striking differences in the descriptors. These could at least in part be linked to the difference in the absolute number of compound extreme events (see Fig. 9.4). But even in places where the number of compound extremes are the same or only show small deviations for both datasets (e.g. in regions along the boarder of Germany and the Netherlands), a difference in the dynamical behavior is visible. The Markov method can therefore be a helpful tool for comparing different observational datasets when one is interested in dynamical properties or in any form of succession of compound extreme events. Possible reasons of these differences are the different density of stations and method of interpolation.

For further comparison, the descriptors from the HYRAS dataset are used as this dataset has a resolution closer to the model resolution, is based on a higher number of stations and the gridding method preserves the occurrence of extreme events (see Chapter 4).

The descriptors for the 7 km ensemble are shown in Fig. 9.6. The top row (Fig. 9.6 a, 1-4) shows the results for the persistence, the second row (b, 1-4) for the recurrence time and the bottom row (c, 1-4) for the entropy. In addition to the ensemble mean (1) , the interquartile range (2:first quartile and 3:third quartile) of the ensemble is depicted. The color scale for the respective descriptors is the same as for the HYRAS results (Fig. 9.5). Figures marked with a 4 in Fig. 9.6 mark grid points where the descriptors calculated from the HYRAS dataset lie within the ensemble spread (mean $\pm\,\sigma$). The ensemble mean shows much less pronounced regional differences than the observations although the individual ensemble members do show noticeable regional structures. The minimum and maximum values of the persistence for the HYRAS domain are smaller than for the observations, but for a large part of the investigation area, the descriptors of the HYRAS observations are within the ensemble spread. The ensemble captures the higher persistences in the southwestern part of the model domain but not in the northeast. The ensemble mean of the recurrence time is lower than that of the observations. For most grid points, the HYRAS recurrence time lies within the ensemble spread, but the spatial correlation between observations and ensemble mean descriptor shows an anti-correlation. These discrepancies can partly originate from the different number of compound extreme events (compare Fig. 9.4 a and 9.7 a) but also from the difference in persistence. The entropy is of similar magnitude as the HYRAS entropy and can capture the regional differences found in the HYRAS entropy, except for the northeastern part of the domain, where the persistence is also too low.

9.2.2 Climate change signal

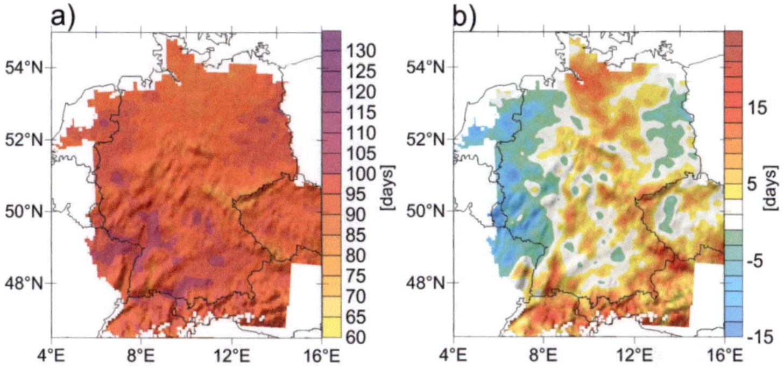

Figure 9.7: Ensemble mean number of compound hot and dry days: a) reference period 1971-2000, b) near future changes, 1971-2000 vs 2021-2050.

The change in the number of compound extreme events between the reference period and the near future is shown in Fig. 9.7 b. There is a decrease in the combined occurrence of hot and dry conditions between 1971-2000 and 2021-2050 in the western part of the investigation area, and an increase which is especially high in the north and around the southern borders of Germany.

The change signal of the descriptors is pictured in Fig. 9.8. As for the reference period, the ensemble mean change signal (1) and the first and third quartile (2+3) are shown for the persistence (first row, Fig. 9.8 a), the recurrence time (second row, Fig. 9.8 b) and the entropy (bottom row, Fig. 9.8 c). The ensemble consistency (1) and the p-level of the significance of the change signal (2) can be found as additional information in Fig. 9.9. The letters correspond to the ones in Fig 9.8 (a = persistence, b= recurrence time, c = entropy). As noted in the last chapter (Chapter 8), the significance may not always be a good measure for change signals of ensembles and a high ensemble consistency is a good indicator of a robust change.

The change signal of the descriptors for compound hot and dry extremes depends strongly on the region and the different ensemble members show different spatial signals of change. The change signal of the persistence depends strongly on the ensemble member. The 25th quantile shows a decrease for the whole investigation area, the 75th quantile an increase for most regions. There are only very few regions where the ensemble consistency is high and the changes are significant at the 5% or 10% significance level (p-value = 0.05/0.10). The recurrence time shows a decrease which is correlated to the decrease in number of events (Fig. 9.7 b). The changes are robust in the Alps, where the ensemble consistency is high (both the 25th and the 75th quantile show this decrease) and changes are mostly significant at the 5%,10% or 20% significance level. The change signal of the entropy is negligibly small for all of the investigation area.
Regions where the persistence shows an increase and the recurrence time a decrease (e.g. some regions in central and northern Germany and small parts of Baden-Wuerttemberg in the southwest) are the ones where the change signal plays the greatest role as this means that the extreme episodes with respect to the new "normal" state become longer and more frequent in addition to a change in mean and variability.

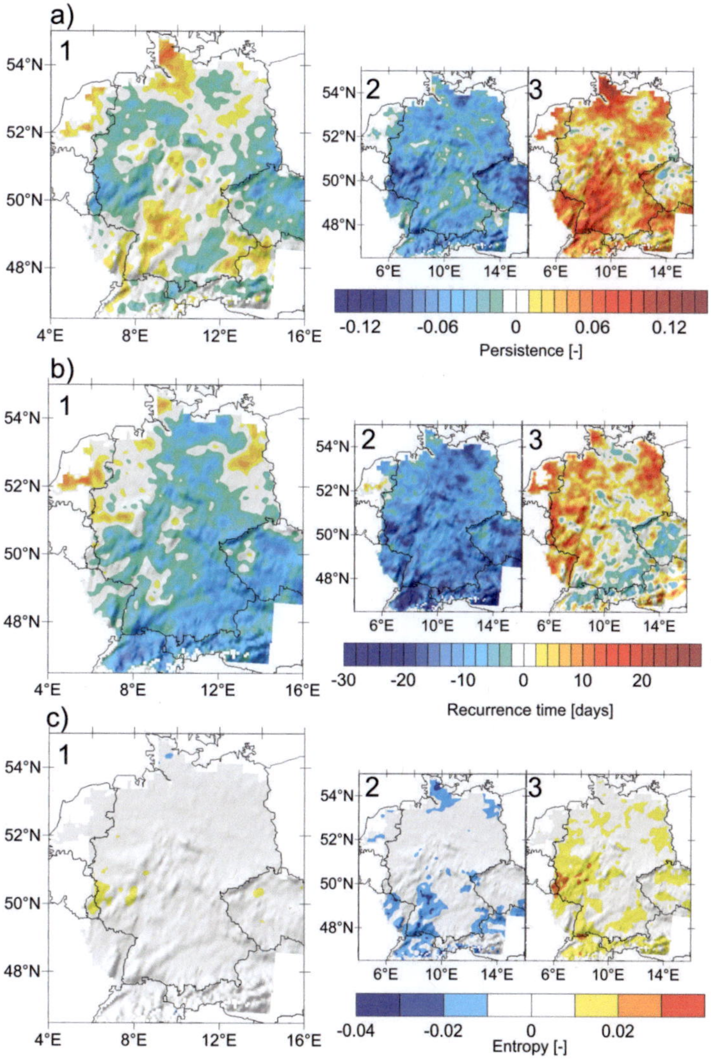

Figure 9.8: Climate change signal of descriptors for compound hot and dry extremes in summer (JJA) between 1971-2000 and 2021-2050. a) persistence, b) recurrence time, c) entropy. 1: ensemble mean change signal, 2: 1st quantile, 3: 2nd quantile.

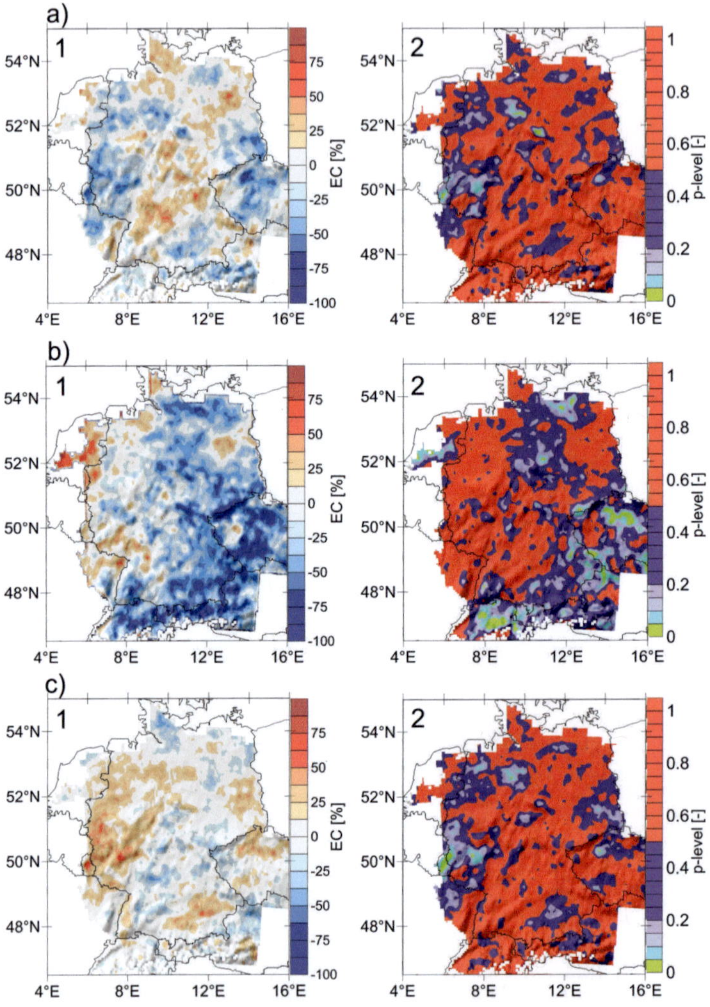

Figure 9.9: Ensemble consistency (left side) and p-level of Wilcoxon test (right side) of the change signal (see Fig. 9.8) of compound hot and dry extremes in summer (JJA). a) persistence, b) recurrence time, c) entropy.

9.3 Cold and wet extremes in winter

9.3.1 Validation

As for compound hot and dry extremes described in the last section, the total number of compound cold and wet extremes in winter (DJF) and the descriptors for the reference period of the two observational datasets are compared before validating the dynamical behavior of the ensemble. The number of compound cold and wet extreme events in winter (DJF) calculated from the HYRAS dataset ranges between 6 and 80 (see Fig. 9.10). Although the number of univariate events are the same for both compound extremes considered in this work (90th vs 10th quantile and 25th vs 75th), as one would expect, cold and wet extremes exhibit a different temporal correlation than heat and drought. The number of compound events is much smaller for cold and wet events in winter. The highest number of events occur in the eastern Alps where the number of absolute compound extremes is also high (see Fig. 7.3). Northeastern Germany shows the lowest number of extremes.

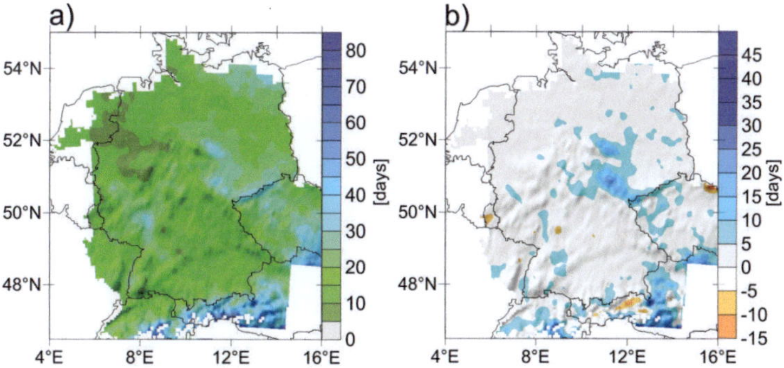

Figure 9.10: Number of compound cold and wet days in winter (DJF), 1971-2000. a) HYRAS, b) HYRAS-E-Obs.

Both observational datasets mostly agree on the number of events, only in the Alpine region and in parts of western Germany the E-Obs dataset shows less compound extreme events. This was already noted for absolute extremes – the E-Obs dataset shows less heavy precipitation events in mountainous regions compared to the HYRAS dataset (see Chapter 7). In Fig. 9.11, the descriptors for the two different observational datasets are compared. For compound cold and dry extremes, the criteria for a stationary transition probability matrix are not met for some of the grid points since the persistence of the compound extreme state is 0. This means that in these regions, which are mostly located in the northwestern part of the investigation area (black colored areas in Fig. 9.11), there is zero probability of an extreme to extreme transition. This is due to the comparably low number of events in this area (5-15 compound extreme days per 30 winters which is equivalent to 1 event every 2 to 6 years). Since the number of days is not very high for all of the investigation area (the maximum value in the HYRAS domain of the ensemble mean is 50 days for the 30-year time period), the persistence is partly correlated to the number of days. Regions with a high number of events also show a higher persistence and a lower recurrence time. The entropy is similar to that of compound hot and dry extremes in summer. It is between 0.22 and 0.24 for most of the domain except for areas with very low persistence (northwestern Germany) where the entropy is very low and the dynamical behavior more predictable. The descriptors of the E-Obs dataset (right side in Fig. 9.11) differ strongly in some regions. In central Eastern Germany the probability of a extreme-extreme transition is about 20% more likely in the HYRAS than in the E-Obs dataset (persistence of 0.32 vs 0.12), the recurrence time is slightly lower for the E-Obs dataset and the entropy only shows small deviations except for the regions in the northwest with 0 persistence. But here, the stationarity conditions are not met. These discrepancies again show that an evaluation of observational datasets with this method can yield valuable additional information if one is not only interested in the number of compound extreme days but also in their succession.

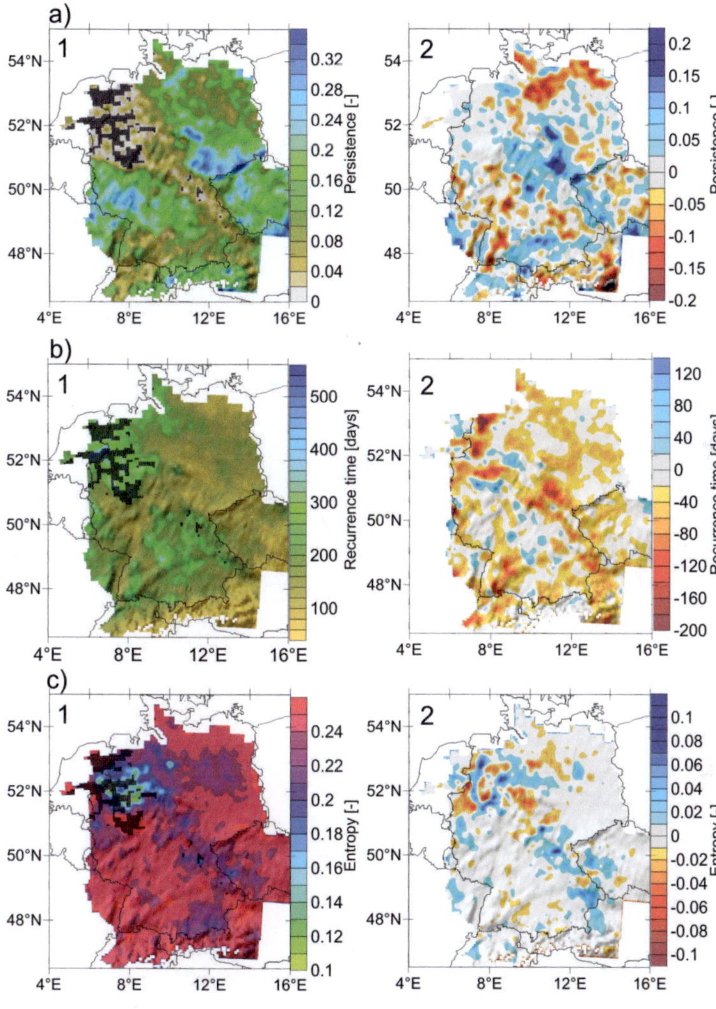

Figure 9.11: Descriptors from Observations for compound cold and wet extremes in winter (DJF) for the reference period, 1971-2000. a) persistence, b) recurrence time, c) entropy. 1) HYRAS descriptors, 2) HYRAS-E-Obs. Grid points where the persistence is 0 and the stationarity criteria violated are marked in black.

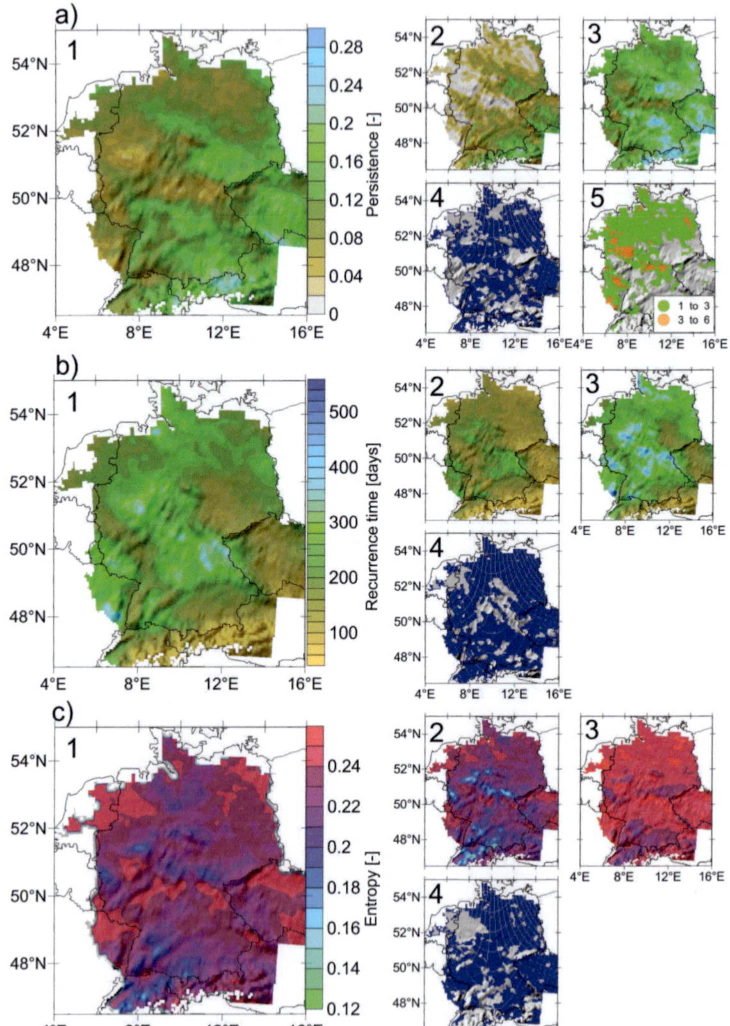

Figure 9.12: Descriptors from the 7 km ensemble for compound cold and wet extremes in winter (DJF) for the reference period, 1971-2000. a) persistence, b) recurrence time, c) entropy. 1: ensemble mean, 2: 1st quantile, 3: 2nd quantile, 4: grid points where HYRAS descriptors are within the ensemble spread (blue). 5: number of ensemble members for which persistence equals zero.

For validation purposes, the HYRAS dataset is again used because of the higher resolution and station density. The number of compound cold and wet extremes in winter (DJF) for the ensemble mean are shown in Fig. 9.13 a. The color scale is the same as for the HYRAS dataset (Fig. 9.10) for better comparison. As for the HYRAS dataset, the highest number of compound extreme events occur in the Alps, and the ensemble mean is able to capture the regional differences exhibited by the observations, however, the overall number of compound cold and wet extremes is slightly underestimated. The ensemble mean as well as the ensemble interquartile range of the descriptors are shown in Fig. 9.12. The color scales are the same as in Fig. 9.11 to enable a direct comparison to the HYRAS descriptors.

The ensemble mean persistence has a similar spatial signal as that of the HYRAS observations but with lower maximum values and a non-zero persistence in the northeastern part of Germany, at least for some ensemble members. The HYRAS descriptors are within the ensemble spread for most grid points. The same can be said for the recurrence time. The correlations between HYRAS descriptors and ensemble mean are 0.42 (persistence) and (recurrence time) respectively. The entropy is also in a similar order of magnitude as for the observations (except for the northeastern part where some of the ensemble members also do not meet the stationarity criteria), and the HYRAS entropy is within the ensemble spread for most grid points but the spatial pattern differs from that of the observations.

9.3.2 Climate change signal

The change signal of the ensemble for the descriptors of compound cold and wet extremes in winter (ensemble mean and first and third quantile) is shown in Fig. 9.14, and the corresponding ensemble consistency and the p-level of significance of the change signal can be found in Fig. 9.15. The first row in both Figs. shows the results for the persistence (a), the middle one for the recurrence time (b) and the bottom row for the entropy (c). The most pronounced changes in persistence are an increase in the northwestern and northeastern part of the model domain where the change is around 0.15, and a decrease in the Czech Republic and southern Germany. These changes are also significant at the 5 % level (p-value = 0.05) and show a high ensemble consistency. In the rest of the investigation area, there are small patches with a consistent and significant change signal, but also large areas where the ensemble members disagree on the sign of the change. The recurrence time mostly shows a decrease which is significant with a high consistency in parts of the south eastern, central and northwestern investigation area. This is correlated with the increase in the number of days as shown in Fig. 9.13b. The highest impact of the changes in dynamical behavior can be found in the southwestern and northwestern parts (consistent and significant increase of persistence and decrease of recurrence time); in these regions the ensemble mean projects longer and more frequent episodes. The entropy shows small changes in the north and the south of which some are significant but not always with a high ensemble consistency.

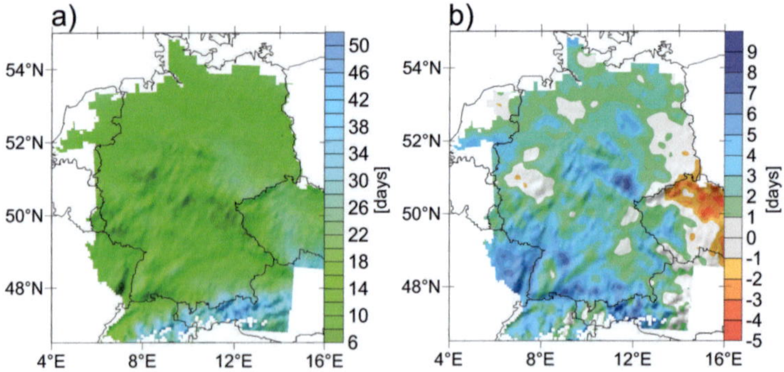

Figure 9.13: Ensemble mean number of compound cold and wet days: a) reference period 1971-2000, winter (DJF), b) near future changes, 1971-2000 vs 2021-2050, winter (DJF).

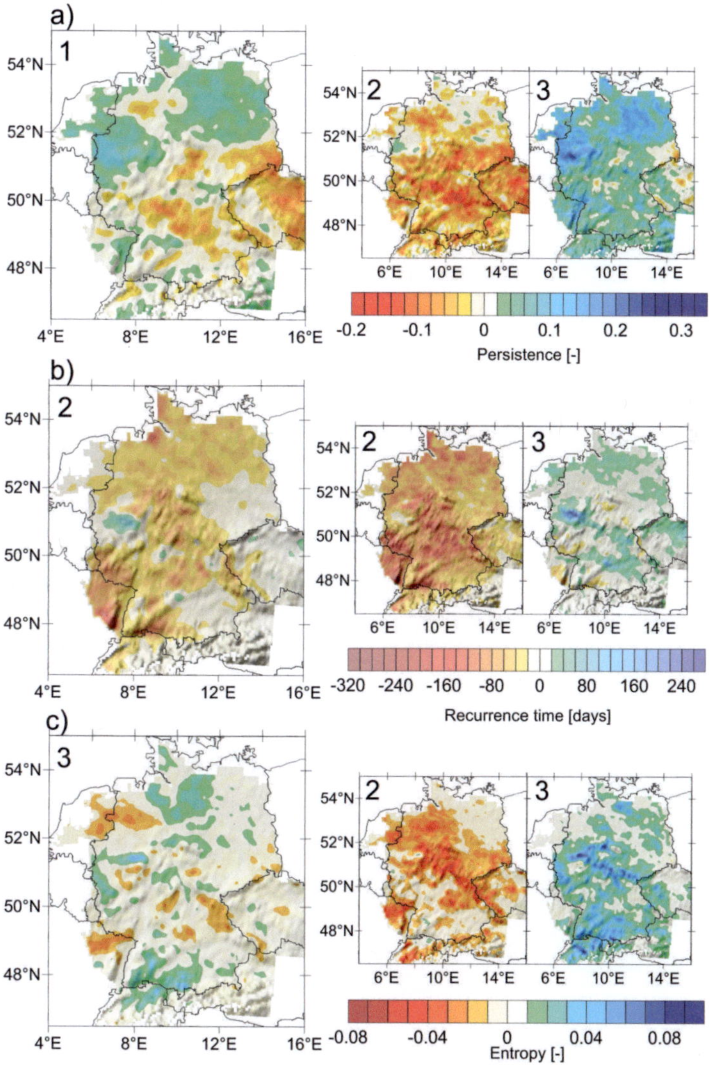

Figure 9.14: Climate change signal of descriptors for compound cold and wet extremes in winter (DJF) between 1971-2000 and 2021-2050.
a) persistence, b) recurrence time, c) entropy. 1: ensemble mean change signal, 2: 1st quantile, 3: 2nd quantile.

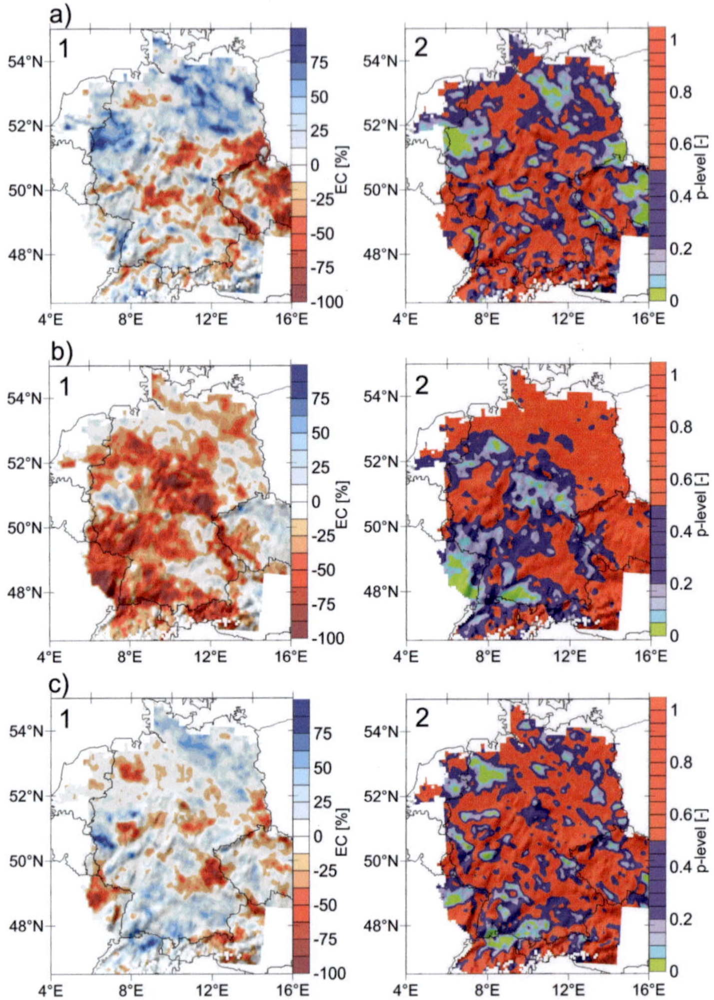

Figure 9.15: Ensemble consistency (left side) and p-level of Wilcoxon test (right side) of the change signal (see Fig. 9.14) for cold and wet extremes in winter (DJF). a) persistence, b) recurrence time, c) entropy.

9.4 Added value of high resolution

For the analysis of added value, the descriptors of the 7 km and 50 km ensemble members were compared to HYRAS observations. The root mean square error (RMSE) of all gridpoints within the HYRAS domain and the spatial correlation between the 7 km and 50 km ensemble members and HYRAS observations are depicted in Fig. 9.16 for hot and dry extremes in summer (JJA) and in Fig. 9.17 for cold and wet extremes in winter (DJF). Boxes show the median and interquartile range of the ensemble, whiskers the ensemble minimum/maximum. Gray bars show the RMSE/correlation of the ensemble mean. The differences between data with 7 km and 50 km resolution are not significant. This fits to the results of the relative extremes (Chapter 8), where

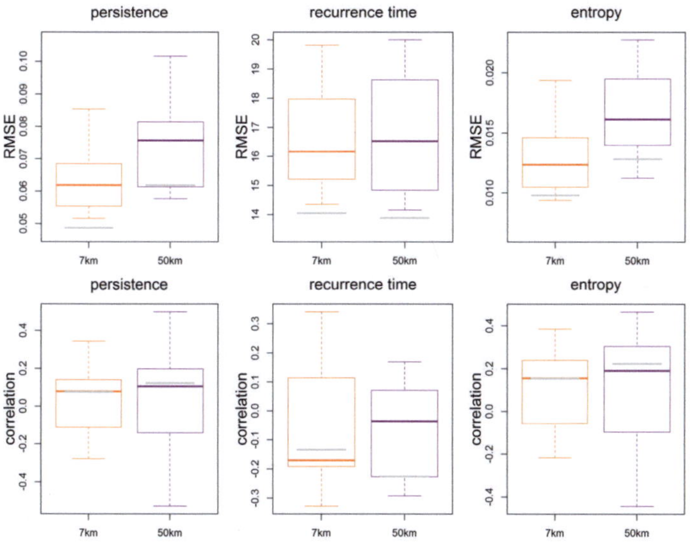

Figure 9.16: Root mean square error (RMSE, top row) and spatial correlation (bottom row) of the 7 km and 50 km ensemble members with respect to HYRAS observations for the HYRAS domain for descriptors of compound for hot and dry extremes in summer (JJA) in the reference period (1971-2000) (see text).

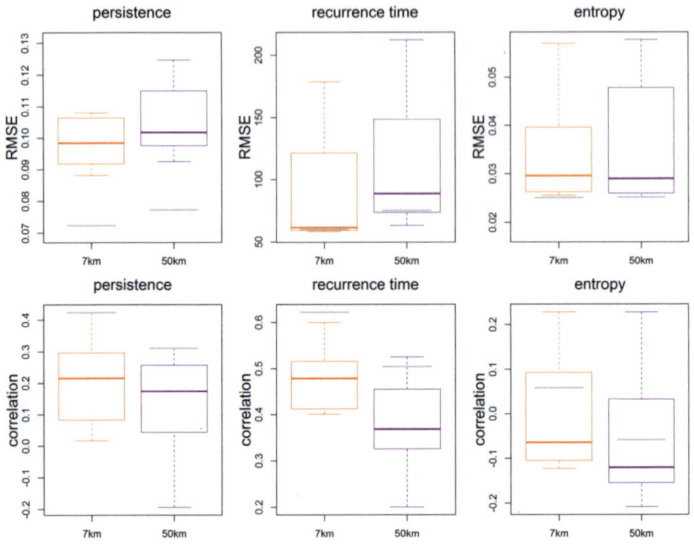

Figure 9.17: Root mean square error (RMSE, top row) and spatial correlation (bottom row) of the 7 km and 50 km ensemble members with respect to HYRAS observations for the HYRAS domain for descriptors of compound for hot and dry extremes in summer (JJA) in the reference period (1971-2000) (see text).

no added value was found either. Typically, the ensemble mean (gray bar) mostly has a smaller RMSE than the individual ensemble members and a better spatial correlation for cold and wet extremes in winter. This underlines that the ensemble mean is a good measure to use.

9.5 Short summary

In this chapter, the dynamical behavior (in terms of persistence, recurrence time and entropy) of hot and dry extremes in summer (JJA) and cold and wet extremes in winter (DJF) was analyzed using a method based on Markov chains which was developed within this work.

An interesting finding are the large discrepancies in some regions between both observational datasets for both kinds of compound extreme events. The difference in persistence is as high as 0.2, which means the probability of the system staying in a compound extreme state differs by up to 20%.

Validation (1971-2000):
Compound hot and dry extremes in summer show the highest observed persistence in eastern France and a small area in northeastern Germany (up to 0.7). In these regions, the recurrence times are also high (up to 130 days). Thus, compound hot and dry episodes have a long duration, but do not occur very frequently. The lowest recurrence times (as low as 45 days, which amounts to two episodes per summer) are found in southeastern Germany and the Czech Republic. The entropy is roughly inversely proportional to the persistence, and with values between 0.19 and 0.25 the occurrence of compound hot and dry extremes shows predictability. Compared to the HYRAS dataset, the 7 km ensemble is able to reproduce the dynamical behavior well in most regions. Compound cold and wet extremes generally yield a lower persistence than cold and wet extremes (all values are below 0.35) and higher recurrence times, which means they do not occur as frequently and the episode lengths are shorter. In northwestern Germany, the persistence is 0. The highest persistences are in Rhineland-Palatinate, Saxony and the Czech Republic, these are paired with comparatively low recurrence times. Thus, these regions yield the highest frequency and duration of compound extreme episodes in the investigation area. The entropy is in a similar range as for hot and dry extremes (except for the areas with zero persistence). The ensemble is able to reproduce the dynamical behavior of the HYRAS observations well for most regions.

Climate change signal (1971-2000 vs 2021-2050):
The change signal between the reference period and the near future largely depends on the region. Significant changes for compound hot and dry ex-

tremes with a high ensemble consistency can be found in the Alps and Czech Republic where the recurrence time shows a decrease by up to 15 days (thus, compound extreme episodes occur more frequently). In the area of Luxemburg, the persistence shows a robust decrease of up to -0.05. In this region, compound hot and dry episodes relative to the new "normal" state will be shorter in the future. Changes in entropy are negligible. For compound cold and wet extremes in winter, the ensemble mean shows a decrease of the recurrence time by up to 200 days in the northern and western part of the investigation area, which is robust in most of the western part leading to a large increase in frequency of cold and wet episodes (in some areas the frequency almost doubles). The persistence shows robust increases by up to 0.09 in the western and northeastern parts of Germany and, thus, episodes of longer duration can be expected in the future. In the Czech Republic, the persistence decreases by up to 0.1 leading to a shortening of episodes relative to the new "normal" state. Changes in entropy (increases and decreases) are found for small patches within the investigation area, some of which are robust but small in magnitude (the highest changes are ≈ 0.03).

Added value of higher resolution (50 km vs 7 km):

A comparison between 7 km and 50 km ensemble yielded no added value for the high resolution simulations.

10 Relation between compound extremes and weather patterns

The occurrence of extreme events is largely related to certain atmospheric circulation patterns (e.g. Fink et al., 2004, 2009; Kunz et al., 2009; Jacobeit et al., 2009). For a better understanding of these dependencies, weather types can be classified and their relation to extreme events studied. In this chapter, this is done for absolute compound extreme events as defined in Chapter 7, since these extremes (at least the hot and dry extremes in summer) show a robust and significant change signal for the near future time period for the whole investigation area. Besides identifying the weather types which lead to (compound) extreme events, the question of whether the change in frequency of extreme events (e.g. the increase of hot and dry events found in Chapter 7) can be attributed to a change in the occurrence of the relevant weather types and whether or not the same weather types are linked to these extremes in the future is investigated. There are numerous methods of weather classification for the European Region (see e.g. Philipp et al., 2010; Schädler and Sasse, 2006). The one used in this work, the objective weather type classification of the German Weather Service (Bissolli and Dittmann, 2001), has the advantage that is was designed for an area that corresponds to the investigation area in this work and it is not dependent on expert judgment, but rather is an algorithm that can be applied to the different climate models of the ensemble. The weather type classification depends on three factors: advection of air masses, cyclonality and humidity of the troposphere. It has been used in the past to study the relationship between extreme events and weather types, e.g. for tornadoes by Bissolli et al. (2007), for hail events by Kapsch et al.

(2012) or temperature and precipitation (and their extremes) by Riediger and Gratzki (2014), or for analyzing the variability of the atmospheric water budget components (Sasse et al., 2013). In the following, the weather type classification scheme is introduced before the above mentioned questions are analyzed for compound hot and dry extremes in summer and cold and wet extremes in winter.

10.1 Objective weather type classification

The objective weather type classification differentiates between 40 possible weather types which are derived by the combination of three meteorological criteria (Bissolli and Dittmann, 2001): the advection of air masses (AA), the cyclonality (near the surface, C_{1000}, and in the mid-troposphere, C_{500}) and the humidity at several levels of the troposphere (H). The weather types depending on these four classes are defined as follows:

$$AAC_{1000}C_{500}H$$

In the following, a short review of the method is given and changes to the original version of Bissolli and Dittmann (2001) are mentioned. The original classification is defined for an area covering the investigation area of this work and, thus, can be easily applied. For the calculation of the different criteria of classification explained below, the grid points are weighted according to Fig. 10.1 when calculating the mean value for the region. The center points (red) are weighed by a factor of three, the grid points around the borders of Germany by a factor of two (blue) and the remaining points by a factor of 1 (yellow). The whole area comprises 21 $\times$ 24 gridpoints.
The grid types of the CLM 50km model domain, which was used for the classification in this work (shown in Fig. 10.1), and the model used in the original classification scheme differ, but the weighting was chosen to be as close to the original weighting scheme as possible.

Different to the original scheme, for which data on 5 pressure levels is used (950, 850, 700, 500 and 300 hPa levels), only 1000, 850, 700 and 500 hPa levels were taken into account in this work due to availability. Other modifications of the classification are mentioned below. The weather types are classified daily at 12 UTC. The three main classification criteria are:

- **AA index - advection in 700 hPa:** XX,NE,SE,SW,NW
  This is based on the zonal and meridional wind components at 700 hPa and identifies the origin and advection of air masses. The large-scale flow direction is derived by dividing the wind rose into four main directions: NE ([0°-90°)), SE ([90°-180°)), SW ([180°-270°)) and NW ([270°-360°)). If more than 2/3 of the grid points show a wind direction in the same sector (note: grid points are weighed according to Fig. 10.1), this is considered as the predominant wind direction, otherwise the wind direction is defined as XX (no prevailing wind direction). This is a slightly modified classification in comparison to Bissolli and Dittmann (2001), where the wind rose is first split into 36 sectors of each 90° extension, for which the number of grid points falling into each sector are counted. If more than 2/3 of the

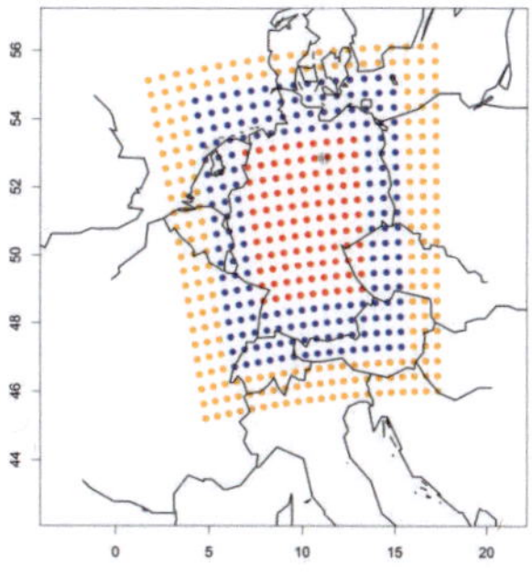

Figure 10.1: Classification area. Colors mark different weights of the grid points: red=3, blue=2,yellow=1.

grid points lie within one of these 36 sectors, the center value defines the main wind direction (for more details see Bissolli and Dittmann (2001)). Kapsch et al. (2012) state that a comparison of both calculation methods for single days yielded more realistic results for the modified classification, which is therefore also used in this work.

- **C_{1000},C_{500} indices: cyclonality in 1000 hPa and 500 hPa:**
  C: cyclonic, A: anticyclonic
  The cyclonality is calculated from the Laplacian of the geopotential ϕ, $\nabla^2\phi$ for each grid point. To derive the cyclonality index, the weighted areal mean is computed, cyclonic conditions show positive values, anticyclonic conditions negative values. The index is calculated for both 1000 hPa and 500 hPa.

- **H index - humidity between 1000 hPa and 500 hPa:** W: wet, D: dry
  The humidity index is calculated from the specific humidity at the 1000, 850, 700 and 500 hPa levels. Wet or dry conditions are classified by checking whether the weighted areal mean of the precipitable water (PW) is above (wet) or below (dry) a long term average for a certain month. PW is calculated as the vertically integrated mixing ratio $r = q_v - \frac{q_v}{0.6222}$ (q_v = specific humidity) for each gridpoint:

  $$PW = -\frac{1}{g}\int_{1000hPa}^{500hPa} r dp \doteq -\frac{1}{g}\sum_{i=1}^{4}\frac{1}{2}(r_{i+1}+r_i)(p_{i+1}-p_i) \quad [10.1]$$

  A monthly long term mean is calculated separately from the reanalysis data and each model for the respective time periods.

For validation purposes, the weather types were calculated from ERA 20C reanalysis. This is a reanalysis product of the ECMWF (www.ecmwf.int/en/research/climate-reanalysis/era-20c), which was an outcome of the ERA-CLIM project (www.era-clim.eu) with a horizontal resolution of $\approx$ 125 km. In addition, weather types for all 12 models of the COSMO-CLM ensemble (see Chapter 3) were calculated for the reference period (1971-2000) and the near future (2021-2050). For the calculation, pressure level data from the 50 km ensemble was used as the weather patterns should not change by further dynamical downscaling since they are determined by the general circulation prescribed by the forcing global climate model. The ERA 20C data was bilinearly interpolated to the COSMO-CLM 50 km model grid before calculation of the weather types. This way the weighting scheme is exactly the same.

10.2 Weather patterns and compound extreme events

The analysis of weather types concentrates on their interrelation with extreme events since these are the main focus of this work. General dependencies of mean temperature and precipitation patterns on the different weather types can be found in Riediger and Gratzki (2014), who used a slightly simplified version of the objective weather type classification. They also analyzed the dependence of univariate temperature and precipitation extremes on the weather type. In the following, it is assessed which weather types lead to type 1 (compound) extreme events (extreme events defined by exceedance of an absolute threshold) as defined in Chapter 7, Table 9.2. This is done by analyzing E-Obs extremes and compound extremes and their relation to weather types derived from the ERA 20C reanalysis data for the reference period, 1971-2000. In a second step, the weather types, for which more than 5% of the compound extremes occur, are compared between reanalysis data and model output, and the change of the dependencies is analyzed for the near future. All extremes are calculated for 7 km model data and E-Obs data bilinearly interpolated to the 7 km grid.

10.2.1 Hot and dry extremes in summer

In Fig. 10.2, the percentage of (compound) hot and dry extreme days related to the respective weather types is shown for extremes calculated from the E-Obs dataset (as in Chapter 7) and weather types calculated from the ERA 20C reanalysis. The values are means over the HYRAS domain – the number of respective weather types prevailing on (compound) extreme days were added up for all grid points and then divided by the sum of all extreme days for all gridpoints in the HYRAS domain. The weather types are ordered by wind direction (XX, NE, SE, SW, NW) and then divided into dry and wet conditions. The third category, the cyclonality (A/C= anticyclonic/cyclonic), is noted below the bars where the first letter marks the cyclonality at 1000 hPa, the second at 500 hPa.

The top graph shows the fraction of hot days for all 40 weather types. The predominant weather types for these extremes are for XX (44%) and SW (43%) advection types with a negative cyclonality index in 500 hPa (94%) and wet conditions (88%), namely SWCAW, SCAAW, XXCAW and XXAAW. Southwesterly flow direction is connected to advection of warm air masses from the lower lattitudes and, thus, induces higher temperatures, while for XX weather types the wind can be very weak and the air masses do not move much which can lead to blocking situations. Anticyclonic conditions in the middle troposphere (500 hPa) related to high pressure systems are known for sunny days. In 1000 hPa, weather types with positive cyclonality index show a slightly higher number of hot days (55%) than those with anticyclonic behavior (45%). This could be due to convective processes induced from the surface due to high radiation, which may induce local low pressure areas near the surface leading to an overall cyclonic behavior for the mean of the investigation area.

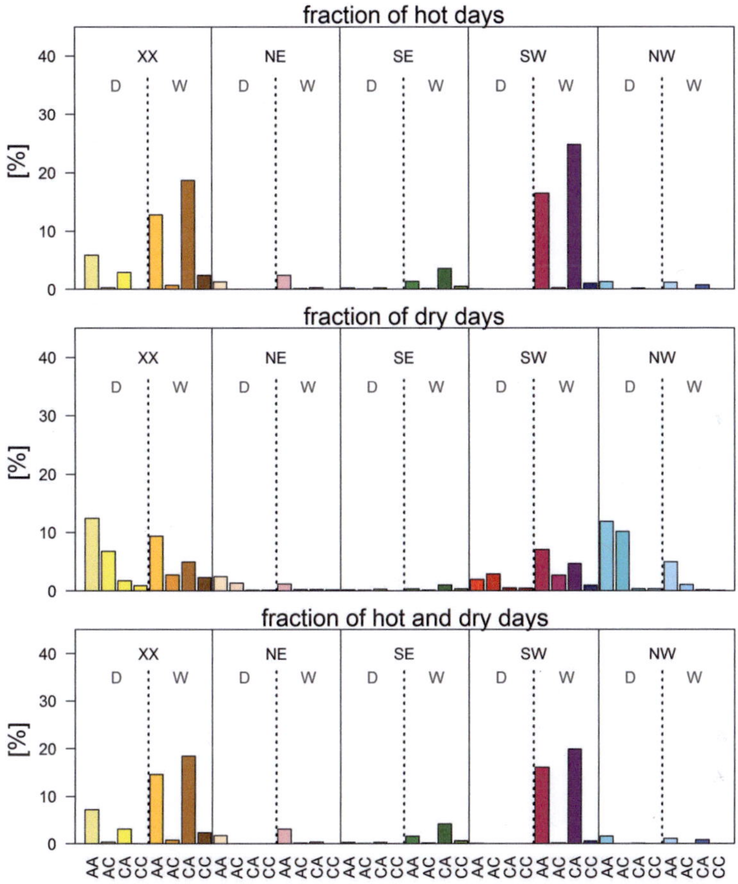

Figure 10.2: Fraction of (compound) hot and dry E-Obs extreme days as defined in Chapter 7 for the different ERA 20C weather types in relation to the total number of extreme days in the HYRAS domain (see Fig. 4.1) for summer (JJA) in the reference period, 1971-2000. Weather types are grouped by advection type (XX, NE, SE, SW, NW), humidity (D/W) and cyclonality in 1000 hPa and 500 hPa (A/C).

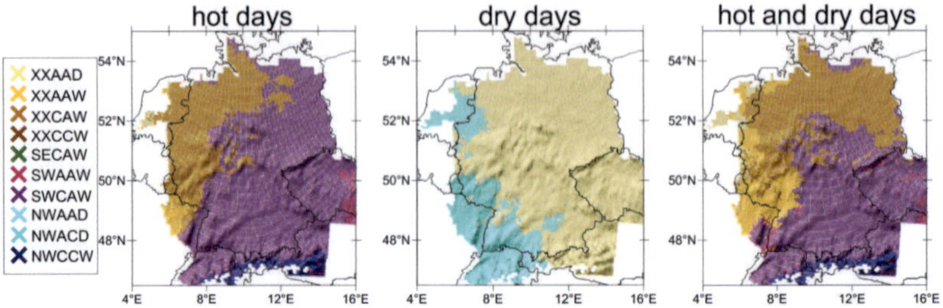

Figure 10.3: Most relevant weather types for (compound) hot and dry extremes in summer (JJA) calculated from the E-Obs dataset and ERA 20C weather types.

Dry days (shown in the middle panel of Fig. 10.2) occur for XX (41%), NW (29%) and SW (22%) advection types. In 66% of the cases these are accompanied by a negative cyclonality index in the middle troposphere (500 hPa) and in 80% of cases the index is negative in the lower troposphere (1000 hPa). As noted above, a negative cyclonality index is a sign of a high pressure system which leads to sunny cloudless skies and therefore low precipitation. The influence of the humidity of the atmosphere is not very high for these extremes, in 55% the air masses are wet and in 45% dry.

The results for hot and dry days fit well to those found by Riediger and Gratzki (2014) for days with mean temperatures above 25°C and days with no precipitation, the same weather types were identified as relevant for those extremes.

The weather types responsible for compound hot and dry days are the same as those for hot days (see lower panel in Fig. 10.2). This is largely due to the fact that the mean number of hot days ($\approx$ 5 per year, see Table 7.2) is about one order of magnitude lower than the number of dry days ($\approx$ 59 per year, see Table 7.2) for the HYRAS domain.

Fig. 10.3 shows the weather types for which the maximum number of extremes occur for each gridpoint for extremes from the E-Obs dataset and weather types from the ERA 20C reanalysis. The dependence on the weather type is very homogeneous for the HYRAS domain. For hot days, SW weather types dominate in the southern and eastern part of the investigation area, whereas in the north and west XX weather types lead to the highest number of hot days, in both cases with a positive cyclonality index for 1000 hPa and a negative one for 500 hPa. In the Alps, most extremes occur under NWCCW weather types, but the number of hot extremes here is low, therefore they do not play an important role in the overall statistics (see Fig 10.2).

The weather type for which the most dry days occur (middle figure) is XXAAD for most of the region followed by anticyclonic NW and XX weather types (not shown in Fig. 10.3). In the western part of the model domain, NWAC weather types additionally play a dominant role. As for the number of compound hot and dry extreme events, the dominant weather types of compound extreme events are largely determined by the occurrence of hot extreme events. Most of the southern and western parts of the investigation area are dominated by SW types, whereas the northern and western parts are dominated by XX weather types. In the Alps, the prevailing weather type is again the NWCCW type but the number of compound hot and dry days here lies below one day per year (see Fig. 7.1).

For analyzing the dependence of the change signal of compound hot and dry days on the change of weather types, all weather types for which at least 5% of compound extreme days occur for E-Obs/ERA 20C or the ensemble (reference and future) are assessed. This yields six weather types, namely XXAAD, XXAAW, XXCAW, SWAAW, SWCAW and NWAAW. Fig.10.4 shows the fraction of compound hot and dry days for the respective weather type in relation to the total number of compound hot and dry extreme days. The first bar for each weather type (o) shows the values for E-Obs/ERA 20C (same values as in Fig. 10.2, bottom pabel), the second and third bar the ensemble mean and ensemble spread (sd $\pm$ σ) for the reference period (r) and near future (f), respectively.

The ensemble is able to capture the influence of the different weather types fairly well. The number of compound extremes occurring on days with XXCAW weather type is underestimated by the ensemble, whereas the number of extremes occurring on days with SWCAW weather type is overestimated. The absolute number of days with XXCAW/SWCAW weather types (for extreme and non extreme days) is also over-/underestimated (not shown) leading to the conclusion that the model is not always able to differentiate between XX and SW wind directions.

One reason for this could be that on those days, the number of gridpoints for the ERA20C reanalysis showing a SW advection type lies slightly below the threshold of 2/3 over which a prevailing wind direction is determined. Another reason could stem from the different resolution of the weather types. However, Kapsch et al. (2012) have shown that the resolution of the respective grids does not play a dominant role, but can be of importance when the cyclonality parameter is close to zero or the humidity is close to the reference value.

The changes between reference period and near future (r and f bars in Fig.10.4) are all below 7%. In general, AA weather types show a higher percentage of compound extreme days in the near future, whereas for CA days there is a decrease. The changes in the distribution between the reference

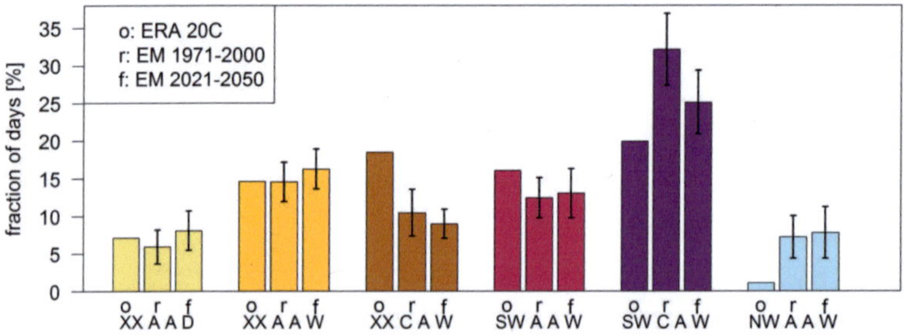

Figure 10.4: Fraction of compound hot and dry extreme days for relevant weather types in relation to the total number of compound extreme days in the HYRAS domain (see Fig. 4.1). Comparison between ERA20C weather types and ensemble mean for summer (JJA) of reference period (1971-2000) and near future (2021-2050). Line segments denote the standard deviation of the ensemble mean.

period and the near future are, however, not very large and not significant except for the SWCAW weather type where a significant decrease of -7% is shown by the ensemble mean. This can be explained by a stronger increase of compound extremes in the western part of the model domain, especially in the Rhine Valley and western France, than in the eastern part (see Fig. 7.2). In the western part of the model domain, the fraction of compound extremes for SW advection types is relatively low and XX weather types are the dominant weather types responsible for compound extremes (see Fig. 10.3).

The absolute number of weather types in summer show a similar tendency as the fraction of hot days - the number of days with AA weather types increases and days with CA weather show a decrease (not shown), but the changes are not significant. Significant changes in the absolute frequency of weather types in summer are only found for weather types which were found not to be relevant for the occurrence of compound hot and dry days (not shown). Therefore, changes in extreme events neither can be clearly attributed to a change in the total number of days of the relevant weather types nor to a change in dependence.

The change in dependence is caused by the magnitude of the change signal of compound extreme events by which the weather types which are predominant in the regions with the biggest change are responsible for a higher fraction of days. A reason for the change in extreme events could be a change in the properties of the air masses that are related to the weather types. With a general increasing temperature trend projected for the near future (e.g. IPCC, 2013), the temperature of the air masses transported by the different advection types is most likely higher in the near future leading to more hot extremes. For example, a temperature increase in Northern Africa or the Atlantic leads to a higher temperature of air masses advected by SW and NW weather types. As the hot extremes are the main prerequisite for the occurrence of compound hot and dry extremes, these would most likely also increase. If air temperatures in the mid troposphere are higher, this could also lead to higher near surface temperatures. This would mean that the properties of the air masses transported by certain weather types could be responsible for the climate change signal rather than the change in frequency of the respective weather types.

10.2.2 Cold and wet extremes in winter

The percentage of (compound) cold and wet extremes related to different weather types for E-Obs extremes and weather types derived from the ERA 20C reanalysis is shown in Fig. 10.5. The procedure is the same as for (compound) hot and dry extremes described in the previous section.

The top figure shows the results for cold extremes. The highest percentage of these extremes occurs on days with no prevailing wind direction (XX, 43%). Cold days also occurr on days with NW and SW advection types (24/19%), whereas for days with NE and SE advection, the percentage is very low. As for hot days, cold days mostly occur for a negative cyclonality index in the middle and lower troposphere (500 hPa and 1000 hPa). The combination of XX advection types and negative cyclonality is known for meridional flow which causes the advection of polar air masses, lower cloudiness and night cooling (Riediger and Gratzki, 2014). NE and SE advection types are also known for severe cold conditions, but their overall number of occurrence is low (8/6%). For more severe cold extremes than the ones defined in this chapter (minimum temperature below 0°C), NE and SE weather types play a more dominant role (Riediger and Gratzki, 2014).There are slightly more cold days for wet than for dry conditions (66% vs. 34%).

The figure in the middle of Fig. 10.5 shows the percentage of wet days attributed to the 40 different weather types. Most wet days occur for NW advection types (50%), followed by SW (30%) and XX (20%) types, the number of extremes for NE and SE advection types is again negligibly small. In general, wet days occur more frequently for weather types with a positive cyclonality index in the mid-troposphere (500 hPa) and for wet conditions.

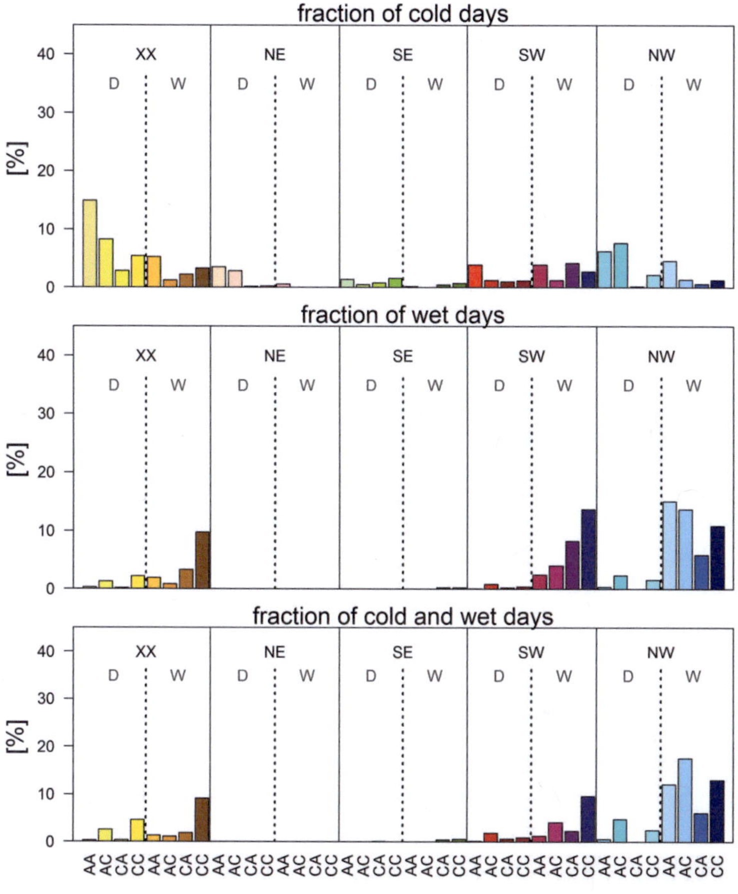

Figure 10.5: Fraction of (compound) cold and wet E-Obs extreme days as defined in Chapter 7 for the different ERA 20C weather types in relation to the total number of extreme days in the HYRAS domain (see Fig. 4.1) for winter (DJF) in the reference period, 1971-2000. Weather types are grouped by advection type (XX, NE, SE, SW, NW), humidity (D/W) and cyclonality in 1000 hPa and 500 hPa (A/C).

NW advection leads to moisture transport from the North Atlantic, SW advection brings warm wet air from the lower latitudes. For the whole HYRAS domain, the mean number of wet days in winter is very low (≈0.4 days per

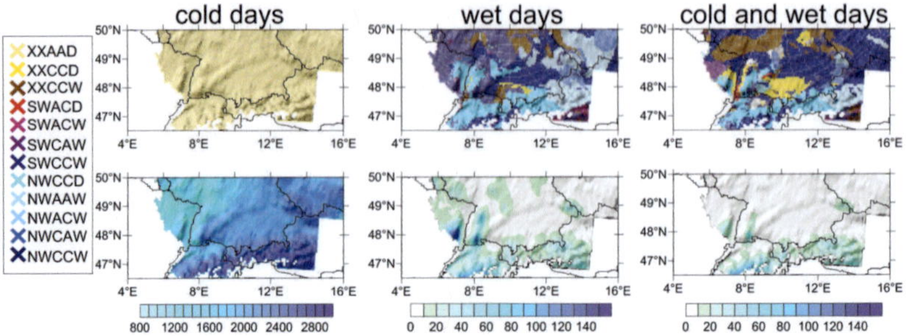

Figure 10.6: Top row: Most relevant weather types for (compound) cold and wet extremes in winter (DJF) calculated from the E-Obs dataset and ERA 20C weather types. Bottom row: total number of (compound) cold and wet extremes in the reference period for the E-Obs data (note the different scale of cold days).

year), and most of these extremes occur where this moist warm air meets orographical barriers like in the Alps, the Vosges or the Black Forest.

The weather types for which compound cold and wet extremes occur are shown in the lower panel of Fig. 10.5. These events are largely dominated by the occurrence of wet days (see Section 7.2) and the weather types, for which most compound extremes occur, are also similar. The lower percentage of compound extremes for SWCC and SWAC weather types in comparison to wet extremes (middle panel) is due to the difference in the spatial pattern of the univariate cold and wet extremes occurring for these different weather types. Cold extremes during SWCC and SWAC weather types mostly occur in the Alps and the eastern part of the model domain, whereas the maximum of wet extremes for SWCC and SWAC types is in the Vosges and the Black Forest (not shown).

The weather types for which the highest number of (compound) cold and wet extremes occur for each grid point are depicted in the top row of Fig. 10.6. Only the southern part of the HYRAS domain is shown as this is the only region where these compound events can be found. As a guide to the eye the absolute number of cold days, wet days and compound cold and wet days are also shown in the bottom row.

The weather type which yields the most cold extreme events for all grid points is the XXAAD type. For this weather type however, no wet extremes occur in the investigation area. NWAA, and AC conditions combined with NW and XX advection types also show a high number of cold extremes in the part of the investigation area shown in Fig. 10.6 (not shown). For wet extremes, the weather types for which the number of extremes is highest, show a high spatial variability. One reason for this is the high spatial variability of precipitation itself. As its occurrence is influenced by orographic features, it depends strongly on the local wind system, and especially in terrain with complex orography this changes for the different advection types. Another reason is the low total number of heavy precipitation events (see middle figure in bottom row in Fig. 10.6). In some areas, less than 5 extremes occur for the whole 30-year time period, making a statistical analysis of the dependencies between weather types and extremes difficult. For those regions which show a higher number of extreme events, two weather types dominate, the NWACW type in the western part of the Alps, the Vosges and the Black Forest and the NWAAW type in the Eastern Alps. These are also the weather types which dominate the occurrence of compound cold and wet extreme events.

For the analysis of the dependence of the change signal of compound cold and wet days on the change of weather types, all weather types for which at least 5% of compound extreme days occur for E-Obs/ERA 20C or the ensemble (reference and future) are assessed. This yields eight weather types, namely XXCCW, SWCCW, NWACD, NWCCD, NWAAW, NWACW, NWCAW and NWCCW.

Fig.10.7 shows the fraction of days with the respective weather types in relation to the total number of compound cold and wet extreme days. The first bar for each weather type (o) shows the values for E-Obs/ERA 20C (same values as in Fig. 10.5), the second and third bar the ensemble mean and ensemble spread (sd $\pm$ σ) for the reference period (r) and near future (f), respectively.

The ensemble attributes too many extreme events to the NWCAW weather type, the fraction for all other wet NW types are overestimated. In sum, the wet NW days match fairly well when omitting the differentiation between the cyclonality. Therefore, a possible reason for the discrepancy could be low cyclonality values, which are thus grouped differently because they are slightly above/below zero even though their absolute values do not differ by much. In this case, the difference in resolution of the ERA 20C data and the 50 km ensemble could be a reason for these discrepancies. Furthermore, the total number of extreme events exhibits a bias between ensemble mean and E-Obs observations (see Fig. 7.3 and Section 7.2). Since the dependence on the weather type varies by region (see Fig. 10.6) and the overall number of extreme events is low, a bias in one region could lead to a significant lower/higher fraction of days with the weather type that is dominant in that region.

In the near future, compound cold and wet extremes occur more likely under NWAC weather types; the increase of these weather types is significant at the 5% level, and the total increase (dry and wet humidity index) is $\approx 5\%$.

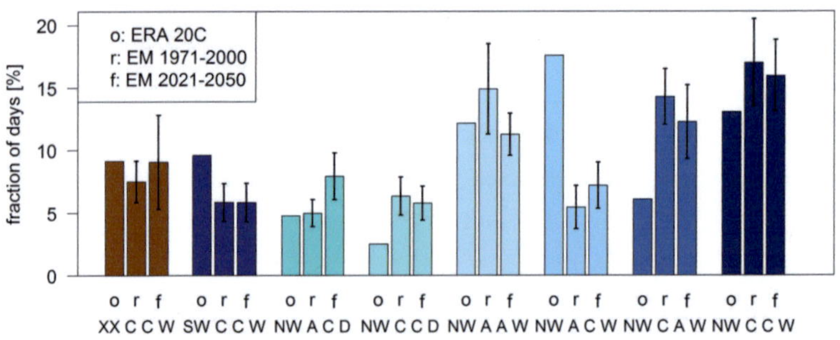

Figure 10.7: Fraction of compound cold and wet extreme days for relevant weather types in relation to the total number of compound extreme days in the HYRAS domain (see Fig. 4.1). Comparison between ERA 20C weather types and ensemble mean for winter (DJF) of reference period (1971-2000) and near future (2021-2050). Line segments denote the standard deviation of the ensemble mean.

The occurrence of compound extremes for NWAAW weather types decreases by about 3.5%. All other changes are not significant. The changes in the absolute number of days with these weather types are also not significant (not shown). As these extremes do not show any significant changes between the reference period and the near future (see Fig. 7.4), it cannot be clearly stated whether or not strong changes in weather types would also induce a change in the number of extreme events. However, there seems to be a change in the dependence of compound extremes on different weather types which is most probably due to a change in the spatial pattern of the extreme events in the future.

10.3 Short summary

In this chapter, the weather types influencing the absolute compound extremes defined in Chapter 7 were analyzed using the objective weather type classification of the German weather service by Bissolli and Dittmann (2001). Clear dependencies between certain weather types and compound extreme events were found for extremes from E-Obs observations and weather types from ERA 20C reanalysis and the ensemble data. About 80% of hot and dry extremes in summer occur on days with south-west or undefined advection types paired with anticyclonic conditions in the mid troposphere and a high precipitable water content. Cold and wet extremes in winter largely depend on northwest advection types with a high precipitable water content with respect to the long term mean. The ensemble is able to reproduce this dependence between weather types and extreme events fairly well. No significant change signal between the absolute number of occurrence of weather types relevant for either type of compound extreme event was found, nor is there a clear change in the dependence of the respective extremes on certain weather types.

A possible reason for the increase of hot and dry compound extreme events found in Chapter 7 could be a change in the properties of the air masses transported by the identified advection types, which may be responsible for the change in extremes.

11 Conclusions

Summary

Compound extremes related to temperature and precipitation, e.g. hot and dry days, bare a high risk potential for society. However, the methods for analyzing these compound extreme events are by far not as manifold and established as for univariate extreme events and in the published literature about extreme events, compound extreme events only male up a small parts. Hence information about these compound events and their future climate change signal is rare. Therefore, in this work, temperature and precipitation extremes, namely hot and dry extremes in summer (JJA) and cold and wet extremes in winter (DJF) were analyzed by three different methods which focus on different aspects of climate change. The first method (type 1 extremes) focuses on absolute extreme events. These are defined by the simultaneous exceedance of temperature and precipitation thresholds on a given day. One sector which is greatly affected by these extremes is agriculture. The second type of extremes are relative extreme events (type 2 extremes) which are derived from indices calculated as standardized anomalies of precipitation and temperature time series while additionally taking the memory effect (the behavior of the preceding days) into account. Extremes by this definition are not necessarily record breaking events but rather a measure of the variability of the climate system. They play an important role especially in regions which are not affected by absolute extremes, but where the ecosystem is nevertheless susceptible to changes in deviations from the mean temperature (e.g. phenology and distribution of species). In addition, the information about the change of variability is relevant as supplementary information to that of absolute extremes. The third method used in this work (type 3 extremes) is a novel method for the assessment of compound extreme events and focuses on the dynamical behavior, i.e. the interplay and succession of extremes. In addition to frequency, duration and intensity, differences in temporal succession between regions and different time periods can be assessed. The method allows to answer the question whether there are changes in this dynamical behavior on top of the changes assessed by the other methods. Thus, the mean and variability of the near future define the new "normal state" and extremes are defined as deviations relative to this and changes in linear trends and variability (assessed by the first two methods) are omitted in the analysis. In addition, the analysis of the predictability of the new state (by calculation of the entropy), enables the detection of differences/changes of the ordering of the system, i.e. if it is more regular or chaotic in different regions/time periods.

For the analysis, an existing high resolution 4-member climate ensemble at 7 km resolution generated with the regional climate model COSMO-CLM was enlarged by dynamically downscaling data of eight global climate models at two nesting stages (50 km and 7 km). These additional members were mostly generated within this work (see Chapter 3). The ensemble data covers the time periods 1971-2000 (reference period) and 2021-2050 (near future). Analysis of the consistency and significance of the change signal are measures for the robustness of changes between these two time periods. The investigation area is central Europe (see Fig.11.1).

Climatological means for the reference period were compared to observational data (see Chapter 4). The comparison yielded a cold and wet bias of the ensemble members which were therefore bias corrected (see Chapter 5). An analysis of the changes in the ensemble mean values of the bias corrected time series between the two time periods showed a significant increase in minimum, maximum and mean temperatures, but no significant changes in precipitation (the majority of ensemble members show a slight decrease in summer and a increase in winter).

The aim of the work was to answer four main questions formulated in the introduction to this work. These focus on compound extreme events, namely their representation by the regional climate model ensemble, their changes in the near future and how the compound extremes depend on different weather patterns. Additionally, the added value of using the high resolution 7 km ensemble compared to the first nesting stage at 50 km was assessed. In the following, these questions are answered, and the results discussed. The corresponding chapters are denoted in brackets.

1. **Can the ensemble of regional climate simulations correctly simulate the statistical occurrence of compound extreme events for a reference time period (1971-2000)?**

   In general the ensemble is able to reproduce the statistical occurrence of compound extremes for the reference period fairly well for most types and aspects of extremes and regions. The ensemble performs better for hot and dry extremes in summer than for cold and wet extremes in winter. (Chapters 7-9)

   With all three methods, the region which yields the most compound hot and dry extreme events in the reference period is the Rhine Valley, where the ensemble mean shows a high number of absolute threshold exceedances (type 1) and relative (type 2) extremes and the persistence is found to be high. A high number of relative (type 2) extremes are also found in the northern part south of Hamburg and in the Alps. The Rhine Valley and the Alps also show a high persistence and, thus, long episodes compared to other regions. In comparison to the observations, the number of hot and dry compound hot and dry type 1 extreme events (absolute extremes) is slightly too high in the southern half of the investigation area, but the deviations are not very large (1-2 days per year). Relative extremes calculated by EDI and EHI (type2) are within the ensemble spread for all but a few grid points in the Northwest and the Markov descriptors (type 3) are also within the ensemble spread for a large fraction of gridpoints.

Cold and wet extremes in winter generally occur less frequently than hot and wet extremes in summer. Besides showing a lower number of absolute (type 1) and relative (type 2) extremes , their persistence is lower and the recurrence time higher (type 3). Both absolute and relative extremes occur in the Alps, the Black Forest and the Vosges. Relative extremes additionally occur in the southern and eastern part of the model domain and in central Germany, where their number, however, shows deviations from the observations. The Markov persistence shows the highest values in the south-eastern part of the model domain, including the Alps, Bavaria, the Czech Republic and parts of Saxony and Brandenburg. In these regions, the recurrence time is also low compared to other regions, which means that type 3 compound episodes events occur often and the episodes are of longer duration. In comparison to observations, cold and wet extremes in winter are represented fairly well for absolute extremes (type 1) and the descriptors of the Markov analysis (type 3), only the observational values for relative extreme events (type 2) lie outside of the ensemble spread for most of the model domain, except for the southern part (Baden-Wuerttemberg and the area around the Alps).

Further investigations and comparisons to multimodel regional climate ensembles (such as e.g. the CORDEX-ensemble) as well as the investigation of the whole model chain (GCM and 2 RCM nesting stages) are necessary to evaluate the source of the respective biases, especially for relative biases.

2. **How will the statistical occurrence of compound extremes change between this reference period (1971 - 2000) and the near future (2021 - 2050) and how robust are the predicted changes? Can re-**

gions be identified which are especially susceptible to the change of extreme events?

Regions affected by climate change of compound extreme events are summarized in Fig. 11.1. All of the investigation area is susceptible to changes in compound hot and dry events which show an increase of absolute (type 1) extremes for the whole investigation area and changes in relative (type 2) extreme events and dynamical behavior (type 3) in the Alps and the eastern part of the model domain. Compound cold and wet events only show robust changes in some regions which are located mostly in the Alps and eastern part as well. These regions are likely to experience an increase in both winter and summer extremes. (Chapters 7 - 9)

How the occurrence of compound extreme events changes in the near future depends on the region, the kind of compound extreme (hot and dry or cold and wet) and the type of the extreme considered (absolute or relative extremes and extremes assessed by the Markov method). If robust changes for all three methods are predicted for a certain area, it is most likely highly susceptible to climate change in the future because the extremes are likely to happen more often, the variability increases and the dynamical behavior changes. Regions which show changes derived by the three methods described in Chapters 7 - 9 are summarized in Fig. 11.1. The images show for which grid points robust changes (shown by more than 75 % of the ensemble members) in the number of absolute compound extreme days (type 1, Chapter 7), relative compound extreme days (type 2, Chapter 8) and/or changes in dynamics properties (type 3, Chapter 9) are predicted by the ensemble.

A change in all three types of hot and dry extremes (black colored areas in Fig. 11.1, left side) is visible in the eastern part of Bavaria, the Czech Republic and its borders with Saxony. These regions can be identified

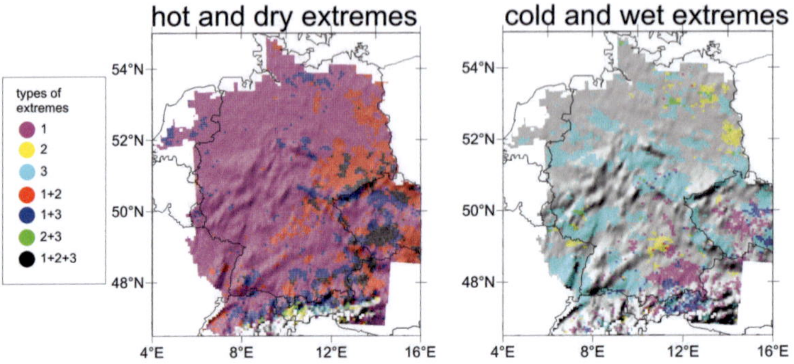

Figure 11.1: Regions where a change occurs between reference period (1971 - 2000) and the near future (2021 - 2050) for compound extremes assessed by three different methods . Type1: absolute extreme events (Chapter 7), type 2: relative extreme events (Chapter 8) and type 3: Markov extremes (Chapter 9). Only changes with an ensemble consistency greater than 50% are considered. Changes can be both negative and positive, gray areas show no changes.

as being highly susceptible to climate change in the future. For the absolute (type 1) number of compound hot extremes, a robust and significant change is projected for all grid points which is highest in the Rhine Valley and Brandenburg (up to 7.5 days). This is in agreement with other studies for univariate extreme events (see Chapter 7). These extremes are dominated by the occurrence of hot extremes and the increase is most likely due to an increasing temperature trend which yields more hot extremes and, thus, also an increase in compound events. In the eastern part of the investigation area, most regions are additionally affected by relative extremes and, thus, a higher variability, i.e., the Alpine region and mountainous regions in the Czech Republic. On top of these changes, the dynamics of succession additionally change in parts of the southern (Alps) and eastern parts of the investigation area. In these regions, the time between compound extreme episodes relative to the new "normal" state decreases by up to 15 days. The change signal for compound cold and wet extremes is of smaller magnitude than for hot and dry extremes and for some regions the change signal is negligibly small. The number of absolute (type 1) and relative (type 2) compound cold and wet extremes only show robust changes in small regions. However, the dynamic properties of these extremes change. The time beweeen episodes relative to the new "normal" state decreases for regions in the western part of the investigation area (for some it is halved) and the episodes area likely to be of longer duration in the northwestern and northeastern parts while in the Czech Republic the results show a shortening of the episode length.

3. **Is there any added value from regional climate simulations at 0.0625 ° (≈ 7 km) resolution in comparison to regional climate simulations at 0.44 ° (≈ 50 km) resolution for the description of compound extreme events?**

   The higher resolution shows an added value for mean values and absolute extremes, mostly by a better spatial correlation of extreme events. For relative extremes, there is no added value. (Chapters 4 , 7 - 9)

   The added value of the different simulations was assessed by comparing mean values and results from the different methods of compound extremes to those of observations. For this, the root mean square error (RMSE) and the spatial correlation were assessed. A summary of the results can be found in Table 11.1. For mean values of temperature and precipitation the 7 km ensemble shows an added value for the spatial correlation. For the temperature means, this is significant for both summer and winter, for the precipitation total, only for winter. The RMSE, however, does not show any added value, but even a small worsening. For compound extreme events, the only added value is found for absolute (type 1) extreme events. Due to lack of availability of temperature maxima/minima data for the 50 km ensemble, only

Table 11.1: Summary of added value of 7 km resolution simulations in comparison to 50 km simulations for mean temperature and precipitation and the three types of compound extreme events. Check marks denote added value. * Added value for type 1 extreme events was only calculated for dry/wet days as temperature data at 50 km was not available.

	JJA	DJF
temperature	✓	✓
precipitation	✓	✓
	hot and dry	cold and wet
type1	- *	✓*
type2	-	-
type3	-	-

the number of dry and wet days, respectively, were compared. These also show a significantly higher correlation between ensemble values and observations. Mean values show no significant difference. For all relative extremes (type 2 and type 3), no significant added value can be found for either extreme.

A downscaling to the higher resolution of 7 km is mostly important when absolute (type 1) extremes are considered on small spatial scales. The relative behavior seems to be governed mostly by the first nesting stage at 50 km, therefore, if one is interested in these extremes (type 2 and 3), further downscaling is not necessarily required.

4. **How are the compound extreme events dependent on different weather patterns and how will these change in the near future?**

   There is a clear dependence of extreme events on certain weather patterns. However, these patterns will not significantly change in the near future. (Chapter 10)

   To study the relation between weather patterns and compound extreme events, the objective weather type classification scheme by Bissolli and Dittmann (2001) was applied to the model data and ERA 20C reanalysis products to derive weather types based on the advection in 700 hPa, the cyclonality in 1000 and 500 hPa and the humidity of the troposphere between 1000 and 5000 hPa. The occurrence of absolute (type 1) compound extremes was set into relation to these weather types. For hot and dry extremes, a clear dependency could be found between Southwest advection type for the south and east of the domain and undefined flow for the western part of the domain with anticyclonic conditions in 500 hPa and a high precipitable water content. Cold and wet extremes in winter largely depend on northwest advection types with a high precipitable water content. However, no significant changes in the occurrence of the relevant weather types could be found and the relevant weather types remain the same in the future. Therefore, the change in extreme events could be induced by a change in the properties of the air masses transported by the different weather types, rather than a change in absolute occurrence.

By answering the questions 1.-4. above, the present work has contributed to the understanding of compound temperature and precipitation extreme events and their occurrence in the future in central Europe. Especially the newly developed analysis of the dynamical behavior reveals information about an aspect which is new to the climate change debate.

Outlook

Possible applications

One possibility of applying the results of this work is the use of the data from the high resolution regional climate simulations as input for impact models, e.g. agricultural models, projections of species development or hydrological models, among others, where regional information, which cannot be provided by the coarsely resolved global climate models, is needed. Through the use of an ensemble of regional climate simulations, the uncertainty of the climate change signal can be included in the impact studies and can be used to identify regions which are most likely impacted by climate change. In addition, the results of Chapters 7-9 can be directly used to identify regions vulnerable to changes in compound extreme events by combining the the climate change signal derived in this work with information about the likely impacts of these changes to the different regions and sectors. For example, the increase in hot and dry extremes in summer for all of Germany will not affect all regions equally, even though the magnitude of the change signal might be the same. Regions, where agriculture is the dominant economic sector, are most probably more affected, although this depends on the crop and its resistance to heat stress and drought. Other examples are health related problems, which are more likely to occur in regions with a high population density or the effects on the ecosystem which depend on the "climatic envelope" of the species living in the respective region (e.g. Walther et al., 2002) and their possibility of migrating to more suitable habitats. For regions identified as vulnerable to compound extreme events, more detailed impact studies using the data of the high resolution study of this work could be conducted.

Further research

Studies of univariate extreme events (e.g. Beniston et al., 2007; Sillmann and Roeckner, 2008; IPCC, 2013) show that other regions of Europe are even more strongly affected by climate change. These regions are most likely also susceptible to compound extreme events. However, information about the regional change signal of compound extreme events with a high spatial resolution is scarce, although it would be of high relevance for designing adaptation measures. Thus, the extension of the analysis to a different/greater region should be a next step. This is underlined by a study accompanying this work (Sedlmeier et al., 2015), which includes an analysis of the 50 km ensemble of this work and focuses on six regions in different parts of Europe. By a Markov Analysis (as in Chapter 9), significant differences between the dynamical behavior of different regions, but also robust changes for some areas (but not the one within the investigation area) were found. A further extension of the study to time periods further in the future would be of high interest in order to answer the question how the dynamical behavior changes for a stronger increase in temperature trend as it is projected for the end of the century (IPCC, 2013).

By further increasing the ensemble size (e.g. by downscaling additional global climate models or using an alternative soil-vegetation-atmosphere model), a broader statistical data basis could be created, which is important for the analysis of such rare events as compound extremes.

This work only includes two types of a large variety of compound extreme events that are relevant to society. The methods used here are generally applicable to compound extremes of two or more variables, although for the relative (type 2) extremes, indices would have to be derived to fit with the meteorological variables of interest. Just as important as the further development of methods to analyze compound extreme events is the information about which compound extreme events are likely to affect society in the future. Not all compounding of extreme events necessarily have a big impact: their potential impact is determined by when, where and how they occur, and a collaboration among different disciplines is necessary. Only if different disciplines work together, the planning of adaptation and mitigation measures can be facilitated to reduce our vulnerability to climate change.

A Appendix

Wilcoxon test

The Wilcoxon test is a non-parametric statistical test for a difference in location of two samples of data. Given two samples, $x_1,...,x_n$ with n values of distribution A and $y_1,...y_m$ with m values of distribution B, the null hypothesis of the Wilcoxon test is:

$$H_o : A = B$$

and the alternative hypothesis:

$$H_1 : A \neq B$$

Under the null hypothesis, the n observations of sample 1 and the m observations of sample 2 are exchangeable, and all come from the same empirical distribution. This test does not depend on absolute values, but rather on the ranks of the samples withing the total number of k=n+m observations. The ranks are assigned by combining the k=n+m observations of both samples and then assigning ranks in ascending order (1 to the smallest, and k to the highest). The rank numbers of the two samples are then summed up where $R_n = \sum_{i=1}^{n} R(x_i)$ are the sums of sample 1 and $R_m = \sum_{i=1}^{m} R(y_i)$ of sample 2. From the rank sum R_n, the U-statistic

$$U = R_n - \frac{n}{2}(n+1)$$

can be calculated. For small sample sizes, tables exist with the critical values of U for rejecting the null-hypothesis for different significance levels (e.g.

Conover, 1999). For larger samples sizes (n and m greater than 10), the critical distributions can be approximated by a normal distribution (Wilks, 2011). The wilcox.stat function of the statistical software package R (R Development Core Team, 2008) returns the U-statistic and the null-distribution probability. Throughout this work, differences are referred to as significant when this probability is lower than 5%, corresponding to a p-level of 0.05, unless stated otherwise.

Bibliography

Allan, R., J. Lindesay, D. Parker, and Coauthors, 1996: *El Nino: Southern oscillation and climatic variability*. CSIRO, 416 pp.

Annan, J., and J. Hargreaves, 2010: Reliability of the CMIP3 ensemble. *Geophys. Res. Lett.*, **37 (2)**, L02 703, doi:10.1029/2009GL041994.

Bartholome, E., and A. Belward, 2005: GLC2000: a new approach to global land cover mapping from Earth observation data. *Int. J. Remote Sens.*, **26 (9)**, 1959–1977.

Bechtold, P., E. Bazile, F. Guichard, P. Mascart, and E. Richard, 2001: A mass-flux convection scheme for regional and global models. *Quart. J. Roy. Meteor. Soc.*, **127 (573)**, 869–886.

Beirlant, J., Y. Goegebeur, J. Segers, and J. Teugels, 2006: *Statistics of extremes: theory and applications*. John Wiley & Sons, 514 pp.

Beniston, M., 2009: Trends in joint quantiles of temperature and precipitation in europe since 1901 and projected for 2100. *Geophys. Res. Lett.*, **36 (7)**, L07 707, doi:10.1029/2008GL037119.

Beniston, M., M. Rebetez, F. Giorgi, and M. Marinucci, 1994: An analysis of regional climate change in switzerland. *Theor. Appl. Climatol.*, **49 (3)**, 135–159.

Beniston, M., and Coauthors, 2007: Future extreme events in European climate: an exploration of regional climate model projections. *Clim. Change*, **81 (1)**, 71–95.

Berg, P., H. Feldmann, and H. J. Panitz, 2012: Bias correction of high resolution regional climate model data. *J. Hydrol.*, **448**, 80–92, doi: {10.1016/j.jhydrol.2012.04.026}.

Berg, P., S. Wagner, H. Kunstmann, and G. Schädler, 2013: High resolution regional climate model simulations for Germany: part I - validation. *Clim. Dyn.*, **40 (1-2)**, 401–414.

Bishop, C. H., and G. Abramowitz, 2013: Climate model dependence and the replicate earth paradigm. *Climate dynamics*, **41 (3-4)**, 885–900.

Bissolli, P., and E. Dittmann, 2001: The objective weather type classification of the German Weather Service and its possibilities of application to environmental and meteorological investigations. *Meteor. Z.*, **10 (4)**, 253–260, doi:10.1127/0941-2948/2001/0010-0253.

Bissolli, P., J. Grieser, N. Dotzek, and M. Welsch, 2007: Tornadoes in germany 1950–2003 and their relation to particular weather conditions. *Glob. Planet. Change*, **57 (1)**, 124–138.

Bruyère, C. L., J. M. Done, G. J. Holland, and S. Fredrick, 2014: Bias corrections of global models for regional climate simulations of high-impact weather. *Clim. Dyn.*, **43 (7-8)**, 1847–1856.

Bukovsky, M., J. Thompson, and L. Mearns, 2013: The Effect of Weighting on the NARCCAP Ensemble Mean. *EGU General Assembly Conference Abstracts*, EGU General Assembly Conference Abstracts, Vol. 15, 2125.

Byun, H.-R., and D. A. Wilhite, 1999: Objective quantification of drought severity and duration. *J. Climate*, **12 (9)**, 2747–2756.

Christensen, J. H., T. R. Carter, M. Rummukainen, and G. Amanatidis, 2007: Evaluating the performance and utility of regional climate models: the prudence project. *Clim. Change*, **81 (1)**, 1–6.

Ciais, P., and Coauthors, 2005: Europe-wide reduction in primary productivity caused by the heat and drought in 2003. *Nature*, **437 (7058)**, 529–533.

Coles, S., J. Bawa, L. Trenner, and P. Dorazio, 2001: *An introduction to statistical modeling of extreme values*. Springer, 209 pp.

Colette, A., R. Vautard, and M. Vrac, 2012: Regional climate downscaling with prior statistical correction of the global climate forcing. *Geophys. Res. Lett.*, **39 (13)**.

Collins, M., 2007: Ensembles and probabilities: a new era in the prediction of climate change. *Philos. Trans. Roy. Soc. London*, **365 (1857)**, 1957–1970, doi:{10.1098/rsta.2007.2068}.

Collins, W., and Coauthors, 2011: Development and evaluation of an Earth-system model–HadGEM2. *Geosc. Model Devel. Disc.*, **4 (2)**, 997–1062.

Conover, W., 1999: *Practical Nonparametric Statistics*. Wiley, New York/Weinheim, 584 pp.

Dai, A., 2013: Increasing drought under global warming in observations and models. *Nature Clim. Change*, **33 (1)**.

Davies, H., 1976: A lateral boundary formulation for multi-level prediction models. *Quart. J. Roy. Meteor. Soc.*, **102 (432)**, 405–418.

Davies, H. C., 1983: Limitations of some common lateral boundary schemes used in regional NWP models. *Mon. Wea. Rev.*, **111 (5)**, 1002–1012.

Déqué, M., and S. Somot, 2010: Weighted frequency distributions express modelling uncertainties in the ensembles regional climate experiments. *Clim. Res.*, **44**, 195209, doi:10.3354/cr00866.

Deser, C., A. Phillips, V. Bourdette, and H. Teng, 2012: Uncertainty in climate change projections: the role of internal variability. *Clim. Dyn.*, **38 (3-4)**, 527–546.

Dickinson, R. E., R. M. Errico, F. Giorgi, and G. T. Bates, 1989: A regional climate model for the western united states. *Clim. Change*, **15 (3)**, 383–422.

Doms, G., and Coauthors, 2011a: A description of the nonhydrostatic regional COSMO model, Part I: Dynamics and Numerics. *Deutscher Wetterdienst, Offenbach, Germany.*

Doms, G., and Coauthors, 2011b: A description of the nonhydrostatic regional COSMO model, Part II: Physical Parameterization. *Deutscher Wetterdienst, Offenbach, Germany.*

Durante, F., and G. Salvadori, 2010: On the construction of multivariate extreme value models via copulas. *Environmetrics*, **21 (2)**, 143–161, doi: 10.1002/env.988.

Evans, J. P., F. Ji, G. Abramowitz, and M. Ekström, 2013: Optimally choosing small ensemble members to produce robust climate simulations. *Environ. Res. Lett.*, **8 (4)**, 044 050.

Feldmann, H., B. Frueh, G. Schaedler, H.-J. Panitz, K. Keuler, D. Jacob, and P. Lorenz, 2008: Evaluation of the precipitation for South-western Germany from high resolution simulations with regional climate models. *Meteor. Z.*, **17 (4)**, 455–465, doi:{10.1127/0941-2948/2008/0295}.

Feldmann, H., G. Schädler, H.-J. Panitz, and C. Kottmeier, 2012: Near future changes of extreme precipitation over complex terrain in Central Europe derived from high resolution RCM ensemble simulations. *Int. J. of Climatol.*, doi:10.1002/joc.3564.

Ferro, C. A., T. E. Jupp, F. H. Lambert, C. Huntingford, and P. M. Cox, 2012: Model complexity versus ensemble size: allocating resources for climate prediction. *Phil. Trans. R. Soc. A*, **370 (1962)**, 1087–1099, doi: 10.1098/rsta.2011.0307.

Feser, F., B. Rockel, H. von Storch, J. Winterfeldt, and M. Zahn, 2011: Regional climate models add value to global model data: a review and selected examples. *Bull. Amer. Meteor. Soc.*, **92 (9)**, 1181–1192.

Fink, A. H., T. Brücher, V. Ermert, A. Krüger, and J. G. Pinto, 2009: The european storm kyrill in january 2007: synoptic evolution, meteorological impacts and some considerations with respect to climate change. *Nat. Hazards Earth Syst. Sci.*, **9 (2)**, 405–423.

Fink, A. H., T. Brücher, A. Krüger, G. C. Leckebusch, J. G. Pinto, and U. Ulbrich, 2004: The 2003 european summer heatwaves and drought–synoptic diagnosis and impacts. *Weather*, **59 (8)**, 209–216.

Fischer, E., and R. Knutti, 2013: Robust projections of combined humidity and temperature extremes. *Nature Climate Change*, **3 (2)**, 126–130.

Fischer, E. M., S. Seneviratne, P. Vidale, D. Lüthi, and C. Schär, 2007: Soil moisture-atmosphere interactions during the 2003 european summer heat wave. *J. Climate*, **20 (20)**, 5081–5099.

Freund, J. A., 1996: Dynamische Entropien und nichtlineare Prozesse mit langreichweitigen Korrelationen. Ph.D. thesis, Humboldt Universität Berlin.

Gallant, A. J., D. J. Karoly, and K. L. Gleason, 2014: Consistent trends in a modified climate extremes index in the united states, europe, and australia. *J. Climate*, **27 (4)**, 1379–1394.

Giorgi, F., 1990: Simulation of regional climate using a limited area model nested in a general circulation model. *J. Climate*, **3 (9)**, 941–963.

Giorgi, F., 2006: Regional climate modeling: Status and perspectives. *Journal de Physique IV (Proceedings)*, EDP sciences, Vol. 139, 101–118.

Giorgi, F., C. Jones, G. R. Asrar, and Coauthors, 2009: Addressing climate information needs at the regional level: the cordex framework. *WMO Bull.*, **58 (3)**, 175–183.

Giorgi, F., and L. O. Mearns, 1999: Introduction to special section: Regional climate modeling revisited. *J. Geophys. Res.*, **104**, 6335–6352.

Gleckler, P. J., K. E. Taylor, and C. Doutriaux, 2008: Performance metrics for climate models. *J. Geophys. Res.*, **113**, D06 104,, doi:10.1029/2007JD008972.

Gudendorf, G., and J. Segers, 2012: Nonparametric estimation of multivariate extreme-value copulas. *J. Statist. Plann. Inference*, **142 (12)**, 3073–3085, doi:{10.1016/j.jspi.2012.05.007}.

Haughton, N., G. Abramowitz, A. Pitman, and S. J. Phipps, 2014: On the generation of climate model ensembles. *Clim. Dyn.*, **43 (7-8)**, 2297–2308.

Haylock, M. R., N. Hofstra, A. M. G. K. Tank, E. J. Klok, P. D. Jones, and M. New, 2008: A European daily high-resolution gridded data set of surface temperature and precipitation for 1950-2006. *J. Geophys. Res.*, **113 (D20)**, doi:{10.1029/2008JD010201}.

Hazeleger, W., and Coauthors, 2010: EC-Earth: a seamless earth-system prediction approach in action. *Bull. Amer. Meteor. Soc.*, **91 (10)**, 1357–1363.

Heino, R., and Coauthors, 1999: Progress in the study of climatic extremes in northern and central europe. *Weather and Climate Extremes*, Springer, 151–181.

Heise, E., 2002: Parametrisierungen. *promet, Meteorologische Fortbildung: Die neue Modellkette des DWD I*, **27**, 130–141.

Hill, M., J. Witman, and H. Caswell, 2004: Markov chain analysis of succession in a rocky subtidal community. *Am. Nat.*, **164 (2)**, E46–E61, doi:{10.1086/422340}.

Hoffmann, H., and T. Rath, 2012: Meteorologically consistent bias correction of climate time series for agricultural models. *Theor. Appl. Climatol.*, **110 (1/2)**, 129 – 141.

Hurrell, J. W., Y. Kushnir, and M. Visbeck, 2001: The north atlantic oscillation. *Science*, **291 (5504)**, 603–605.

IPCC, 2010: *Meeting Report of the Intergovernmental Panel on Climate Change Expert Meeting on Assessing and Combining Multi Model Climate Projections*. IPCC Working Group I Technical Support Unit, University of Bern, Bern, Switzerland, 117 pp.

IPCC, 2012: *Managing the Risks of Extreme Events and Disasters to Advance Climate Change Adaptation. A Special Report of Working Groups I and II of the Intergovernmental Panel on Climate Change*. Cambridge University Press,Cambridge, UK, and New York, NY, USA, 582 pp.

IPCC, 2013: *Climate Change 2013: The Physical Science Basis. Contribution of Working Group I to the Fifth Assessment Report of the Intergovernmental Panel on Climate Change 2013*. Cambridge University Press, Cambridge, UK, and New York, NY, USA, 1535 pp.

Jacob, D., and Coauthors, 2014: EURO-CORDEX: new high-resolution climate change projections for European impact research. *Reg. Environ. Change*, **14 (2)**, 563–578.

Jacobeit, J., J. Rathmann, A. Philipp, and P. D. Jones, 2009: Central european precipitation and temperature extremes in relation to large-scale atmospheric circulation types. *Meteor. Z.*, **18 (4)**, 397–410.

Jones, R., J. Murphy, and M. Noguer, 1995: Simulation of climate change over europe using a nested regional-climate model. i: Assessment of control climate, including sensitivity to location of lateral boundaries. *Quart. J. Roy. Meteor. Soc.*, **121 (526)**, 1413–1449.

Kain, J. S., 1993: Convective parameterization for mesoscale models: The Kain-Fritsch scheme. *The representation of cumulus convection in numerical models, Meteor. Monogr*, **24 (46)**, 165–170.

Kapsch, M.-L., M. Kunz, R. Vitolo, and T. Economou, 2012: Long-term trends of hail-related weather types in an ensemble of regional climate models using a bayesian approach. *J. Geophys. Res*, **117**, D15 107, doi: 10.1029/2011JD017185.

Katz, R. W., and B. G. Brown, 1992: Extreme events in a changing climate: variability is more important than averages. *Climatic change*, **21 (3)**, 289–302.

Keetch, J. J., G. M. Byram, and Coauthors, 1968: A drought index for forest fire control,USDA Forest Service Research Paper SE-38, Southeastern Forest Experiment Sta- tion, Asheville, NC. 33.

Kjellström, E., F. Boberg, M. Castro, J. H. Christensen, G. Nikulin, and E. Sánchez, 2010: Daily and monthly temperature and precipitation statistics as performance indicators for regional climate models. *Clim. Res.*, **44 (2)**, 135–150, doi:10.3354/cr00932.

Kunz, M., J. Sander, and C. Kottmeier, 2009: Recent trends of thunderstorm and hailstorm frequency and their relation to atmospheric characteristics in southwest germany. *Int. J. of Climatol.*, **29 (15)**, 2283–2297.

Laprise, R., 2008: Regional climate modelling. *J. Comput. Phys.*, **227 (7)**, 3641–3666.

Lavalle, C., and Coauthors, 2009: Climate change in europe. 3. impact on agriculture and forestry. a review. *Agron. Sustainable Dev.*, **29 (3)**, 433–446.

McGregor, J., 1997: Regional climate modelling. *Meteor. Atmos. Phys.*, **63 (1-2)**, 105–117.

Mearns, L. O., and Coauthors, 2012: The North American regional climate change assessment program: overview of phase I results. *Bull. Amer. Meteor. Soc.*, **93 (9)**, 1337–1362.

Meehl, G. A., C. Covey, K. E. Taylor, T. Delworth, R. J. Stouffer, M. Latif, B. McAvaney, and J. F. Mitchell, 2007: The WCRP CMIP3 multimodel dataset: A new era in climate change research. *Bull. Amer. Meteor. Soc.*, **88 (9)**, 1383–1394.

Mellor, G. L., and T. Yamada, 1982: Development of a turbulence closure model for geophysical fluid problems. *Rev. Geophys.*, **20 (4)**, 851–875.

Mieruch, S., S. Noel, H. Bovensmann, J. P. Burrows, and J. A. Freund, 2010: Markov chain analysis of regional climates. *Nonlin. Proc. Geoph.*, **17 (6)**, 651–661, doi:{10.5194/npg-17-651-2010}.

Mikosch, T., 2006: Copulas: Tales and facts. *Extremes*, **9 (1)**, 3–20.

Mizielinski, M., and Coauthors, 2014: High-resolution global climate modelling: the upscale project, a large-simulation campaign. *Geoscientific Model Development*, **7 (4)**, 1629–1640.

Moss, R. H., and Coauthors, 2010: The next generation of scenarios for climate change research and assessment. *Nature*, **463 (7282)**, 747–756.

Nakicenovic, N., and R. Swart, 2000: Special report on emissions scenarios. *Special Report on Emissions Scenarios, Edited by Nebojsa Nakicenovic and Robert Swart, pp. 612. ISBN 0521804930. Cambridge, UK: Cambridge University Press, July 2000.*

Norris, J. R., 1998: *Markov chains*. 2008, Cambridge university press.

Philipp, A., and Coauthors, 2010: Cost733cat–a database of weather and circulation type classifications. *Phys. Chem. Earth*, **35 (9)**, 360–373.

R Development Core Team, 2008: *R: A Language and Environment for Statistical Computing*. Vienna, Austria, R Foundation for Statistical Computing, URL http://www.R-project.org, ISBN 3-900051-07-0.

Rauthe, M., H. Steiner, U. Riediger, A. Mazurkiewicz, and A. Gratzki, 2013: A Central European precipitation climatology–Part I: Generation and validation of a high-resolution gridded daily data set (HYRAS). *Meteor. Z.*, **22 (3)**, 235–256.

Reichler, T., and J. Kim, 2008: How well do coupled models simulate today's climate? *Bull. Amer. Meteor. Soc.*, **89 (3)**, 303–311.

Renard, B., and M. Lang, 2007: Use of a gaussian copula for multivariate extreme value analysis: some case studies in hydrology. *Adv. Water Resour.*, **30 (4)**, 897–912.

Riediger, U., and A. Gratzki, 2014: Future weather types and their influence on mean and extreme climate indices for precipitation and temperature in central europe. *Meteor. Z.*

Ritter, B., and J.-F. Geleyn, 1992: A comprehensive radiation scheme for numerical weather prediction models with potential applications in climate simulations. *Mon. Wea. Rev.*, **120 (2)**, 303–325.

Roeckner, E., and Coauthors, 2003: The atmospheric general circulation model ECHAM 5. PART I: Model description. Tech. rep., Max-Planck-Institut für Meteorologie, Hamburg, Germany.

Sasse, R., and G. Schädler, 2014: Generation of regional climate ensembles using Atmospheric Forcing Shifting. *Int. J. Climatol.*, **34 (7)**, 2205–2217, doi:{10.1002/joc.3831}.

Sasse, R., G. Schädler, and C. Kottmeier, 2013: The regional atmospheric water budget over southwestern Germany under different synoptic conditions. *J. Hydrometeorol.*, **14 (1)**, 69–84.

Schädler, G., and R. Sasse, 2006: Analysis of the connection between precipitation and synoptic scale processes in the eastern mediterranean using self-organizing maps. *Meteor. Z.*, **15 (3)**, 273–278.

Schädler, G., and Coauthors, 2012: Flood hazard in a changing climate. Tech. Rep. http://www.cedim.de/download/Flood_Hazards_in_a_Changing_Climate.pdf, Center for Disaster Management and Risk Reduction Technology (CEDIM).

Schär, C., P. L. Vidale, D. Lüthi, C. Frei, C. Häberli, M. A. Liniger, and C. Appenzeller, 2004: The role of increasing temperature variability in european summer heatwaves. *Nature*, **427 (6972)**, 332–336.

Schättler, U., and U. Blahak, 2015: A description of the nonhydrostatic regional COSMO model, Part IV: Preprocessing: Initial and Boundary Data for the COSMO-Model. *Deutscher Wetterdienst, Offenbach, Germany*.

Schmidli, J., and C. Frei, 2005: Trends of heavy precipitation and wet and dry spells in switzerland during the 20th century. *Int. J. of Climatol.*, **25 (6)**, 753–771.

Schoelzel, C., P. Friederichs, and Coauthors, 2008: Multivariate non-normally distributed random variables in climate research–introduction to the copula approach. *Nonlin. Proc. Geoph.*, **15 (5)**, 761–772.

Schrodin, E., and E. Heise, 2002: A new multi-layer soil model. *COSMO newsletter*, **2**, 149–151.

Scinocca, J., N. McFarlane, M. Lazare, J. Li, D. Plummer, and Coauthors, 2008: The CCCma third generation AGCM and its extension into the middle atmosphere. *Atmos. Chem. Phys. Discuss*, **8 (2)**, 7883–7930.

Sedlmeier, K., S. Mieruch, and G. Schädler, 2015: Compound extremes in a changing climate - a markov chain approach. *to be submitted*.

Sedlmeier, K., and G. Schädler, 2014: Ensembles hoch aufgelöster regionaler Klimasimulationen zur Analyse regionaler Klimaänderungen in BaWü und ihre Auswirkungen. Tech. rep., LUBW Landesanstalt für Umwelt, Messungen und Naturschutz Baden - Württemberg.

Shannon, C., 1948: A mathematical theory of communication. *Syst. Tech. J.*, **27**, 623–656.

Sillmann, J., and E. Roeckner, 2008: Indices for extreme events in projections of anthropogenic climate change. *Clim. Change*, **86 (1-2)**, 83–104.

Smiatek, G., B. Rockel, and U. Schättler, 2008: Time invariant data preprocessor for the climate version of the cosmo model (cosmo-clm). *Meteor. Z.*, **17 (4)**, 395–405.

Smodydzin, L., 2004: Theoretische und numerische Untersuchungen zur Konvektionsparametrisierung in einem Wettervorhersagemodell. M.S. thesis, Universität Bonn.

Stevens, B., and Coauthors, 2013: Atmospheric component of the MPI-M Earth System Model: ECHAM6. *J. Adv. Model. Earth Syst.*, **5 (2)**, 146–172.

Taylor, K. E., R. J. Stouffer, and G. A. Meehl, 2012: An overview of cmip5 and the experiment design. *Bull. Amer. Meteor. Soc.*, **93 (4)**, 485–498.

Tebaldi, C., and R. Knutti, 2007: The use of the multi-model ensemble in probabilistic climate projections. *Phil. Trans. R. Soc. A*, **365 (1857)**, 2053–2075, doi:10.1098/rsta.2007.2076.

Tebaldi, C., and B. Sansó, 2009: Joint projections of temperature and precipitation change from multiple climate models: a hierarchical bayesian approach. *J. Roy. Stat. Soc.*, **172 (1)**, 83–106.

Teutschbein, C., and J. Seibert, 2012: Bias correction of regional climate model simulations for hydrological climate-change impact studies: Review and evaluation of different methods. *J. Hydrol.*, **456**, 12–29.

Thuiller, W., S. Lavorel, M. B. Araújo, M. T. Sykes, and I. C. Prentice, 2005: Climate change threats to plant diversity in europe. *Proc. Natl. Acad. Sci. U. S. A.*, **102 (23)**, 8245–8250.

Tiedtke, M., 1993: Representation of clouds in large-scale models. *Mon. Wea. Rev.*, **121 (11)**, 3040–3061.

Trewin, B., O. Baddour, and H. Kontongomde, 2007: *The role of climatological normals in a changing climate*. World Meteorological Organization.

Uppala, S. M., and Coauthors, 2005: The ERA-40 re-analysis. *Quart. J. Roy. Meteor. Soc.*, **131 (612)**, 2961–3012.

Van der Linden, P., and J. Mitchell, Eds., 2009: *ENSEMBLES: Climate Change and its Impacts: Summary of research and results from the ENSEMBLES project*. Met Office Hadley Centre, Exeter, UK.

Voldoire, A., and Coauthors, 2013: The CNRM-CM5.1 global climate model: description and basic evaluation. *Clim. Dyn.*, **40 (9-10)**, 2091–2121.

von Storch, H., and F. Zwiers, 2013: Testing ensembles of climate change scenarios for statistical significance. *Clim. Change*, **117 (1-2)**, 1–9.

Wagner, S., P. Berg, G. Schädler, and H. Kunstmann, 2013: High resolution regional climate model simulations for Germany: Part II - projected climate changes. *Clim. Dyn.*, **40 (1-2)**, 415–427.

Walther, G.-R., and Coauthors, 2002: Ecological responses to recent climate change. *Nature*, **416 (6879)**, 389–395.

Wang, Y., L. R. Leung, J. L. McGREGOR, D.-K. Lee, W.-C. Wang, Y. Ding, and F. Kimura, 2004: Regional climate modeling: progress, challenges, and prospects. *J. Meteor. Soc. Japan*, **82 (6)**, 1599–1628.

Weigel, A. P., R. Knutti, M. A. Liniger, and C. Appenzeller, 2010: Risks of model weighting in multimodel climate projections. *J. Climate*, **23 (15)**, 4175 – 4191.

Weigel, A. P., M. A. Liniger, and C. Appenzeller, 2008: Can multi-model combination really enhance the prediction skill of probabilistic ensemble forecasts? *Quart. J. Roy. Meteor. Soc.*, **134 (630)**, 241–260, doi:{10.1002/qj.210}.

Wicker, L. J., and W. C. Skamarock, 2002: Time-splitting methods for elastic models using forward time schemes. *Mon. Wea. Rev.*, **130 (8)**, 2088–2097.

Wilby, R. L., T. Wigley, D. Conway, P. Jones, B. Hewitson, J. Main, and D. Wilks, 1998: Statistical downscaling of general circulation model output: a comparison of methods. *Water Resour. Res.*, **34 (11)**, 2995–3008.

Wilks, D. S., 2011: *Statistical methods in the atmospheric sciences, 3rd ed.* Academic press, San Diego, CA, 676 pp.

Will, A., and M. Wold, 2009: Comparison of COSMO-CLM results with CM-SAF products: radiation components ToA, at the surface and cloud properties IOP VS Study 16. Tech. rep., Climate Monitoring Satellite Application Facility (CMSAF) at DWD, Offenbach.

Yan, J., and Coauthors, 2007: Enjoy the joy of copulas: with a package copula. *J. Stat. Softw.*, **21 (4)**, 1–21.

Wissenschaftliche Berichte des Instituts für Meteorologie und Klimaforschung des Karlsruher Instituts für Technologie (0179-5619)

Bisher erschienen:

Nr. 1: *Fiedler, F. / Prenosil, T.*
Das MESOKLIP-Experiment. (Mesoskaliges Klimaprogramm im Oberrheintal). August 1980

Nr. 2: *Tangermann-Dlugi, G.*
Numerische Simulationen atmosphärischer Grenzschichtströmungen über langgestreckten mesoskaligen Hügelketten bei neutraler thermischer Schichtung. August 1982

Nr. 3: *Witte, N.*
Ein numerisches Modell des Wärmehaushalts fließender Gewässer unter Berücksichtigung thermischer Eingriffe. Dezember 1982

Nr. 4: *Fiedler, F. / Höschele, K. (Hrsg.)*
Prof. Dr. Max Diem zum 70. Geburtstag. Februar 1983 (vergriffen)

Nr. 5: *Adrian, G.*
Ein Initialisierungsverfahren für numerische mesoskalige Strömungs modelle. Juli 1985

Nr. 6: *Dorwarth, G.*
Numerische Berechnung des Druckiderstandes typischer Geländeformen. Januar 1986

Nr. 7: *Vogel, B.; Adrian, G. / Fiedler, F.*
MESOKLIP-Analysen der meteorologischen Beobachtungen von mesoskaligen Phänomenen im Oberrheingraben. November 1987

Nr. 8: *Hugelmann, C.-P.*
Differenzenverfahren zur Behandlung der Advektion. Februar 1988

Nr. 9: *Hafner, T.*
Experimentelle Untersuchung zum Druckwiderstand der Alpen. April 1988

Nr. 10: *Corsmeier, U.*
Analyse turbulenter Bewegungsvorgänge in der maritimen atmosphärischen Grenzschicht. Mai 1988

Nr. 11: *Walk, O. / Wieringa, J.(eds)*
Tsumeb Studies of the Tropical Boundary-Layer Climate. Juli 1988

Nr. 12: *Degrazia, G. A.*
Anwendung von Ähnlichkeitsverfahren auf die turbulente Diffusion in der konvektiven und stabilen Grenzschicht. Januar 1989

Nr. 13: *Schädler, G.*
Numerische Simulationen zur Wechselwirkung zwischen Landoberflächen und atmophärischer Grenzschicht. November 1990

Nr. 14: *Heldt, K.*
Untersuchungen zur Überströmung eines mikroskaligen Hindernisses in der Atmosphäre. Juli 1991

Nr. 15: *Vogel, H.*
Verteilungen reaktiver Luftbeimengungen im Lee einer Stadt – Numerische Untersuchungen der relevanten Prozesse. Juli 1991

Nr. 16: *Höschele, K.(ed.)*
Planning Applications of Urban and Building Climatology – Proceedings of the IFHP / CIB-Symposium Berlin, October 14-15, 1991. März 1992

Nr. 17: *Frank, H. P.*
Grenzschichtstruktur in Fronten. März 1992

Nr. 18: *Müller, A.*
Parallelisierung numerischer Verfahren zur Beschreibung von Ausbreitungs- und chemischen Umwandlungsprozessen in der atmosphärischen Grenzschicht. Februar 1996

Nr. 19: *Lenz, C.-J.*
Energieumsetzungen an der Erdoberfläche in gegliedertem Gelände. Juni 1996

Nr. 20: *Schwartz, A.*
Numerische Simulationen zur Massenbilanz chemisch reaktiver Substanzen im mesoskaligen Bereich. November 1996

Nr. 21: *Beheng, K. D.*
Professor Dr. Franz Fiedler zum 60. Geburtstag. Januar 1998

Nr. 22: *Niemann, V.*
Numerische Simulation turbulenter Scherströmungen mit einem Kaskadenmodell. April 1998

Nr. 23: *Koßmann, M.*
Einfluß orographisch induzierter Transportprozesse auf die Struktur der atmosphärischen Grenzschicht und die Verteilung von Spurengasen. April 1998

Nr. 24: *Baldauf, M.*
Die effektive Rauhigkeit über komplexem Gelände – Ein Störungstheoretischer Ansatz. Juni 1998

Nr. 25: *Noppel, H.*
Untersuchung des vertikalen Wärmetransports durch die Hangwindzirkulation auf regionaler Skala. Dezember 1999

Nr. 26: *Kuntze, K.*
Vertikaler Austausch und chemische Umwandlung von Spurenstoffen über topographisch gegliedertem Gelände. Oktober 2001

Nr. 27: *Wilms-Grabe, W.*
Vierdimensionale Datenassimilation als Methode zur Kopplung zweier verschiedenskaliger meteorologischer Modellsysteme. Oktober 2001

Nr. 28: *Grabe, F.*
Simulation der Wechselwirkung zwischen Atmosphäre, Vegetation und Erdoberfläche bei Verwendung unterschiedlicher Parametrisierungsansätze. Januar 2002

Nr. 29: *Riemer, N.*
Numerische Simulationen zur Wirkung des Aerosols auf die troposphärische Chemie und die Sichtweite. Mai 2002

Nr. 30: *Braun, F. J.*
Mesoskalige Modellierung der Bodenhydrologie.
Dezember 2002

Nr. 31: *Kunz, M.*
Simulation von Starkniederschlägen mit langer Andauer über Mittelgebirgen. März 2003

Nr. 32: *Bäumer, D.*
Transport und chemische Umwandlung von Luftschadstoffen im Nahbereich von Autobahnen – numerische Simulationen.
Juni 2003

Nr. 33: *Barthlott, C.*
Kohärente Wirbelstrukturen in der atmosphärischen Grenzschicht. Juni 2003

Nr. 34: *Wieser, A.*
Messung turbulenter Spurengasflüsse vom Flugzeug aus.
Januar 2005

Nr. 35: *Blahak, U.*
Analyse des Extinktionseffektes bei Niederschlagsmessungen mit einem C-Band Radar anhand von Simulation und Messung.
Februar 2005

Nr. 36: *Bertram, I.*
Bestimmung der Wasser- und Eismasse hochreichender konvektiver Wolken anhand von Radardaten, Modellergebnissen und konzeptioneller Betrachtungen. Mai 2005

Nr. 37: *Schmoeckel, J.*
Orographischer Einfluss auf die Strömung abgeleitet aus Sturmschäden im Schwarzwald während des Orkans „Lothar". Mai 2006

Nr. 38: *Schmitt, C.*
Interannual Variability in Antarctic Sea Ice Motion: Interannuelle Variabilität antarktischer Meereis-Drift.
Mai 2006

Nr. 39: *Hasel, M.*
Strukturmerkmale und Modelldarstellung der Konvektion über Mittelgebirgen. Juli 2006

Ab Band 40 erscheinen die Wissenschaftlichen Berichte des Instituts für Meteorologie und Klimaforschung bei KIT Scientific Publishing (ISSN 0179-5619). Die Bände sind unter www.ksp.kit.edu als PDF frei verfügbar oder als Druckausgabe bestellbar.

Nr. 40: *Lux, R.*
Modellsimulationen zur Strömungsverstärkung von orographischen Grundstrukturen bei Sturmsituationen
ISBN 978-3-86644-140-8

Nr. 41: *Straub, W.*
Der Einfluss von Gebirgswellen auf die Initiierung und Entwicklung konvektiver Wolken
ISBN 978-3-86644-226-9

Nr. 42: *Meißner, C.*
High-resolution sensitivity studies with the regional climate model COSMO-CLM
ISBN 978-3-86644-228-3

Nr. 43: *Höpfner, M.*
Charakterisierung polarer stratosphärischer Wolken mittels hochauflösender Infrarotspektroskopie
ISBN 978-3-86644-294-8

Nr. 44: *Rings, J.*
Monitoring the water content evolution of dikes
ISBN 978-3-86644-321-1

Nr. 45: *Riemer, M.*
Außertropische Umwandlung tropischer Wirbelstürme: Einfluss auf das Strömungsmuster in den mittleren Breiten
ISBN 978-3-86644-766-0

Nr. 46: Nicht erschienen.

Nr. 47: Nicht erschienen.

Nr. 48: Nicht erschienen.

Nr. 49: *Peters, T.*
Ableitung einer Beziehung zwischen der Radarreflektivität, der Niederschlagsrate und weiteren aus Radardaten abgeleiteten Parametern unter Verwendung von Methoden der multivariaten Statistik
ISBN 978-3-86644-323-5

Nr. 50: *Khodayar Pardo, S.*
High-resolution analysis of the initiation of deep convection forced by boundary-layer processes
ISBN 978-3-86644-770-7

Nr. 51: *Träumner, K.*
Einmischprozesse am Oberrand der konvektiven atmosphärischen Grenzschicht
ISBN 978-3-86644-771-4

Nr. 52: Nicht erschienen.

Nr. 53: *Lundgren, K.*
Direct Radiative Effects of Sea Salt on the Regional Scale
ISBN 978-3-86644-773-8

Nr. 54: *Sasse, R.*
Analyse des regionalen atmosphärischen Wasserhaushalts unter Verwendung von COSMO-Simulationen und GPS-Beobachtungen
ISBN 978-3-86644-774-5

Nr. 55: *Grenzhäuser, J.*
Entwicklung neuartiger Mess- und Auswertungsstrategien für ein scannendes Wolkenradar und deren Anwendungsbereiche
ISBN 978-3-86644-775-2

Nr. 56: *Grams, C.*
Quantification of the downstream impact of extratropical transition for Typhoon Jangmi and other case studies
ISBN 978-3-86644-776-9

Nr. 57: *Keller, J.*
Diagnosing the Downstream Impact of Extratropical Transition Using Multimodel Operational Ensemble Prediction Systems
ISBN 978-3-86644-984-8

Nr. 58: *Mohr, S.*
Änderung des Gewitter- und Hagelpotentials im Klimawandel
ISBN 978-3-86644-994-7

Nr. 59: *Puskeiler, M.*
Radarbasierte Analyse der Hagelgefährdung in Deutschland
ISBN 978-3-7315-0028-5

Nr. 60: *Zeng, Y.*
Efficient Radar Forward Operator for Operational Data Assimilation within the COSMO-model
ISBN 978-3-7315-0128-2

Nr. 61: Nicht erschienen.

Nr. 62: *Jerger, D.*
Radar Forward Operator for Verification of Cloud Resolving Simulations within the COSMO Model
ISBN 978-3-7315-0172-5

Nr. 63: *Maurer, V.*
Vorhersagbarkeit konvektiver Niederschläge : Hochauflösende Ensemblesimulationen für Westafrika
ISBN 978-3-7315-0189-3

Nr. 64: *Stawiarski, C.*
Optimizing Dual-Doppler Lidar Measurements of Surface Layer Coherent Structures with Large-Eddy Simulations
ISBN 978-3-7315-0197-8

Nr. 65: *Mahlke, H.*
Mechanismen der Auslösung hochreichender Konvektion im südwestdeutschen Mittelgebirgsraum
ISBN 978-3-7315-0203-6

Nr. 66: *Fosser, G.*
Precipitation statistics from regional climate model at resolutions relevant for soil erosion
ISBN 978-3-7315-0227-2

Nr. 67: *Adler, B.*
Boundary-Layer Processes Producing Mesoscale Water-Vapour Variability over a Mountainous Island
ISBN 978-3-7315-0247-0

Nr. 68: *Kraut, I.*
Separating the Aerosol Effect in Case of a „Medicane"
ISBN 978-3-7315-0405-4

Nr. 69: *Breil, M.*
Einfluss der Boden-Vegetation-Atmosphären Wechselwirkungen auf die dekadische Vorhersagbarkeit des Westafrikanischen Monsuns
ISBN 978-3-7315-0420-7

Nr. 70: *Lott, F. F.*
Wind Systems in the Dead Sea and Footprints in Seismic Records
ISBN 978-3-7315-0596-9

Nr. 71: *Rieger, D.*
Der Einfluss von natürlichem Aerosol auf Wolken über Mitteleuropa
ISBN 978-3-7315-0672-0

Nr. 72: *Loewe, K.*
Arctic mixed-phase clouds. Macro- and microphysical insights with a numerical model
ISBN 978-3-7315-0686-7

Nr. 73: *Piper, D. A.*
Untersuchung der Gewitteraktivität und der relevanten großräumigen Steuerungsmechanismen über Mittel- und Westeuropa
ISBN 978-3-7315-0701-7

Nr. 74: *Metzger, J.*
Wind Systems and Energy Balance in the Dead Sea Valley
ISBN 978-3-7315-0699-7

Nr. 75: *Deetz, K.*
Assessing the Aerosol Impact on Southern West African Clouds and Atmospheric Dynamics
ISBN 978-3-7315-0744-4

Nr. 76: *Ehmele, F. T.*
Stochastische Simulation großflächiger, hochwasserrelevanter Niederschlagsereignisse.
ISBN 978-3-7315-0761-1